QUAMITRY ~ TH  ᛕ

THE GEOMETRIC COMPRESSION INDEX OF ELEMENTS

DECODED FROM THE PAST ~ DRAFTED FOR THE PRESENT ~ ENGINEERED FOR THE FUTURE.

BY KEITH MOUNTJOY

Quamitry ~ The GCI Codex

Author: Keith Mountjoy. Published by Quamitry Labs.

Published by Quamitry Labs.

For more information, visit

https://www.quamitry.com/

ISBN: 979-8-9940693-0-1

Cover art was created by

iamabbas.com

Printed in the United States of America

First Edition V.1

~A textbook from the future, deciphered from the past~

Part of the Quamitry Codex Series

Table of Contents

Phase Chapters

Each phase functions as a mini book with its own opening scatter, data table, Codex reflections, and closing fragment.

PHASE	TITLE	PAGE RANGE	DOMINANT FOLD TYPE	EPOCH THEME
1	The Primordial Memory (H–Ne)	46-68	Shellfold · Refold	Containment and Reflection
2	The Conductive Epoch (Na–Ca)	69-91	Chainfold · Lockfold	Polarity Learns to Move
3	The Magnetic Epoch (Sc–Zn)	92-115	Lockfold · Crossfold	Fields of Memory
4	The Reflective Epoch (Ga–Kr)	116-131	Lockfold · Refold	Resonance Organization
5	The Resonant Expansion (Rb–Xe)	132-171	Crossfold · Chainfold	Overload and Release
6	The Transitional Epoch (Cs-Lu)	172-209	Chainfold · Fracture Fold	Structural Experimentation
7	The Complex Epoch (Hf–Hg)	210-232	Lockfold · Crossfold	Layered Resonance
8	The Collapse Epoch (Tl–Rn)	233–247	Fracture Fold	Over-compression and Memory Loss
9	The Actinide Epoch (Fr–Lr)	248-285	Fracture Fold · Crossfold	Transience and Rebirth
10	The Transactinide Threshold (Rf-Og)	286-319	Lockfold · Chainfold	Geometry Reawakens

Preface: The Codex as a Language

There are moments when discovery does not arrive with noise, but with recognition ~ a quiet sense that the universe has been whispering all along, waiting for us to learn its grammar. The Geometric Compression Index (GCI) is born from that recognition. It is not a chart of atoms; it is a translation of geometry itself ~ a way to read the hidden pulse of matter.

What we have called "element" is not a substance, but a decision made by resonance ~ to hold, to breathe, to remember. Within every fold of geometry is instruction: how to form, how to endure, how to release. The GCI gives language to those instructions. It is the lexicon of endurance and collapse, the mirror through which structure finally speaks.

Humanity has always measured the visible: the weight of things, the distance between them, the heat of their motion. But here, in this Codex, we begin to measure the invisible decisions that matter makes ~ the moment geometry chooses memory over motion, form over freedom.

Through the framework of Quamitry, geometry ceases to be silent. We discover that stability is not stillness but rhythm; that matter is a song of compression and release, recorded in folds of resonance. The GCI allows us to hear it.

This is more than a scientific system ~ it is a new literacy.

If we can read the resonance of matter, we can rewrite the material itself. We can grow architecture instead of building it, design propulsion that follows harmonic law rather than combustion, and craft energy systems that harmonize instead of consume.

The Codex opens a bridge between physics and purpose ~ between the mathematics of endurance and the poetry of becoming.

Through it, we may finally learn how to move among the stars not by conquering them, but by resonating with them.

Prefatory Statement: The Quamitric Foundation

Before the table of elements, before the lattice of metrics, there is pattern.
The GCI ~ the Geometric Compression Index ~ is built upon that pattern's logic.

That logic is called Quamitry.

Quamitry is not a belief or a branch of science.
It is the *grammar* through which all sciences speak ~
the study of how resonance folds into form and how geometry remembers motion.

In the pages ahead, each element is treated not as mere matter, but as a folded event ~
a moment where resonance became instruction.

The GCI Codex measures those folds.
It is the practical mathematics of Quamitry ~
the place where geometry, energy, and memory finally speak the same language.

Resonance folds geometry. Geometry remembers resonance.
This is the only law beneath all others.

Section 2 ~ The Purpose of the Geometric Compression Index

The GCI was created to answer a question older than chemistry itself: why does matter remember its shape?

Where classical physics measures what matter does, the GCI measures what geometry decides.

It quantifies the silent conversation between resonance and compression ~ the dialogue that gives every atom its endurance, its temper, its pulse.

At its core, the Index converts observation into understanding.

Each element's behavior ~ its reactivity, stability, color, or magnetism ~ is not random; it is the visible trace of how well its internal resonance remains coherent under pressure.

The GCI expresses this coherence as a single signature: a number on the Codex scale from 0 to 9.

High values reveal architectures of deep memory and balance; lower ones mark geometries that trade endurance for transformation.

Both are necessary. The universe survives through the interplay of the steadfast and the volatile.

Traditional atomic tables stop at weight, charge, and configuration.

The GCI extends beneath them, mapping the intent of geometry itself.

It tracks how much energy ~ Geometric Oscillation Energy, or GOE ~ is stored, redirected, or released as matter folds upon itself.

It tells us not just what an element is, but how it became possible.

By quantifying these hidden dynamics, the Codex allows us to compare not only substances, but behaviors of resonance.

We begin to see iron and carbon not as separate materials, but as related solutions to the same question of stability.

We see that biology, magnetism, and even time itself are written in the same alphabet of compression.

The purpose of the GCI, then, is translation.

It gives humanity a common tongue between geometry and energy ~ between the silence of form and the voice of resonance.

Through it, the sciences can converge; engineers can design by harmonic law rather than trial; and explorers may one day tune propulsion fields to the natural frequency of space itself.

To measure a thing is to know its limits.

To measure geometry by its resonance is to glimpse its potential.

That is the work of the Geometric Compression Index ~ the first mirror by which matter may at last know itself.

Section 3 ~ The Core Mathematical Framework

The Geometric Compression Index (GCI) transforms the invisible architecture of resonance into a measurable, reproducible scale.

It quantifies how efficiently a structure holds, redirects, and releases Geometric Oscillation Energy (GOE) ~ the fundamental pulse of existence.

Each element's geometry can be understood as a network of harmonic decisions: how tightly to fold, how freely to move, and how much of its internal resonance to retain before release.

The GCI expresses these decisions mathematically through five resonance axes (REP, DBI, ERF, RRP, FSI) with spatial weighting by MRI inside the Deca-Axis framework, producing a singular resonance value between 0 and 9 on the Codex scale.

Fundamental relation with Deca normalization

Let the internal resonance integrity be the mean of the five resonance axes:

$\bar{R}$ = (REP + DBI + ERF + RRP + FSI) $\div$ 5.

Let the spatiotemporal weight be **W = MRI $\div$ 10** (a 0-1 factor reflecting external coherence).

Raw score: **GCI_raw = $\bar{R}$ $\times$ W.**

Normalize to the 0-9 Codex scale: **GCI = 9 $\times$ GCI_raw** (clamped to [0, 9]).

Reading. High GCI emerges when internal resonance is balanced and retained (large $\bar{R}$) and when the mirrorfield supports coherence (moderate MRI). FSI adds the timebeat of the fold ~ geometries that saturate filaments sustain coherence longer and score higher when the mirrorfield does not over-burden them.

In simple terms:

- High GCI (7-9): geometry of endurance ~ low entropy ~ strong harmonic closure.
- Moderate GCI (5-6): adaptive geometry ~ balance between flow and retention.

- Low GCI (0-4): reactive or transient geometry ~ fast energy exchange ~ high entropy.

The mathematics of the GCI therefore express the moral of the geometry itself: how much energy can be remembered before it must be released. It is the bridge between structure and motion ~ the first equation that allows matter to be read not by its mass, but by its memory.

SECTION 3.5 ~ THE DECA-AXIS RESONANT MAP

Every fold in the Lattice is a dialogue between geometry and memory. The Deca-Axis System completes that dialogue with ten measurable expressions of resonance ~ from compression and retention to spatial coherence and temporal saturation. These axes do not compete; they circulate.

The Ten Axes

GCI ~ Geometric Compression Index ~ Form strength under coherent compression.
GOE Load ~ Geometric Oscillation Energy available prior to compression.
RRP ~ Resonant Release Potential ~ readiness of outflow without collapse.
ERF ~ Energy Retention Factor ~ containment of resonance after excitation.
DBI ~ Density Balance Index ~ symmetry of internal pressure and packing.
REP ~ Resonant Efficiency Potential ~ throughput of oscillation into structure.
FTR ~ Failure Threshold Rating ~ tolerance prior to resonance fracture.
Collapse Vector ~ qualitative mode of release at failure.
MRI ~ Mirrorfield Requirement Index ~ spatial coherence ~ how much reflection is needed.
FSI ~ Filament Saturation Index ~ temporal coherence ~ how long resonance remains.

Spatiotemporal envelope: MRI governs *where* a fold holds in space; FSI governs *how long* it endures in time. Together they complete the circuit of Quamitry.

Section 4 ~ The Axes of Quamitric Measurement

The Quamitry Geometric Compression Index (GCI) Codex defines a Deca-Axis system ~ ten measurable axes that describe how matter folds Geometric Oscillation Energy (GOE) into stable form. Each axis traces a different facet of this process ~ from internal resonance balance and retention to external field dependency, to the precise way a fold collapses when pushed beyond its limit.

In practice, five resonance axes articulate the element's internal behavior ~ REP, DBI, ERF, RRP, FSI ~ while MRI expresses spatial coherence and the remaining structural axes complete the system. Together they describe how resonance moves, stabilizes, reflects, and releases through the lattice of matter. Each axis can be measured numerically, yet each also carries a philosophical aspect ~ a behavior of being. In the Codex, these axes are not abstractions; they are the pulse of geometry made legible.

---

1. REP ~ Resonant Efficiency Potential

Symbol: **REP**
Domain: **0.00 – 1.00**
Function: Measures the efficiency with which GOE (Geometric Oscillation Energy) stabilizes into usable resonance.

High REP signifies folds that channel GOE directly into coherence ~ efficient geometries that resist entropy. Low REP indicates turbulent or wasteful resonance, where energy disperses faster than structure can retain it. REP defines the "muscle" of a fold ~ the power with which geometry holds itself together.

At its philosophical level, REP is the will of matter to remember motion. Every increase in REP represents a step toward self-sustaining form, where energy learns to inhabit its own boundaries.

Classification: Dynamic Behavior Indicator ~ internal cohesion.

---

2. DBI ~ Density Balance Index

Symbol: **DBI**
Domain: **0.00 – 1.00**
Function: Quantifies the equilibrium between compression (inward) and expansion (outward) within a structure.

DBI expresses symmetry ~ the centeredness of a fold. When DBI approaches 1.00, the element's geometry achieves perfect internal balance; compression and expansion coexist in harmonic rhythm. Lower DBI reflects imbalance: either excessive density (leading to brittleness) or excessive release (leading to volatility).

Philosophically, DBI is the axis of grace. It shows that endurance does not require rigidity, but rhythm. A perfectly balanced fold breathes ~ compressing just enough to remember, releasing just enough to remain alive.

Classification: Dynamic Behavior Indicator ~ structural harmony.

---

3. ERF ~ Energy Retention Factor

Symbol: **ERF**
Domain: **0.00 – 1.00**
Function: Determines how much GOE remains within the fold after each resonance cycle.

High ERF means long-term structural stability ~ energy remains stored within the geometry, producing matter that persists through time. Low ERF corresponds to fleeting resonance states, where energy quickly departs and the structure reverts to potential.

Philosophically, ERF is the axis of memory. It is how geometry learns to hold experience. Each increment in ERF is another moment of time captured within form, another breath of resonance refusing to fade.

Classification: Dynamic Behavior Indicator ~ temporal energy retention.

---

4. RRP ~ Resonant Release Potential

Symbol: RRP Domain: 0.00 – 1.00 Function: Measures how effectively energy can be released or transferred without geometric collapse.

High RRP values signify controlled release ~ geometries capable of transferring energy harmonically (such as conductors or catalytic elements). Low RRP geometries tend to store energy until sudden collapse or ignition. RRP defines how a structure communicates its resonance ~ the grace of letting go.

Philosophically, RRP is the axis of generosity. It shows that stability without release becomes stagnation. Every stable geometry must know how to share its motion without losing itself.

Classification: Dynamic Behavior Indicator ~ energy exchange behavior.

5. MRI ~ Mirrorfield Requirement Index

Symbol: **MRI**
Domain: **0 – 10**
Function: Quantifies the strength of the external field required to sustain internal coherence.

MRI represents the "cost" of reflection ~ how much ambient resonance (or external field symmetry) an element needs to remain stable. Low MRI elements are self-contained, holding their shape with minimal environmental feedback. High MRI elements rely heavily on external resonance ~ they are cooperative geometries, thriving within fields and bonds.

Philosophically, MRI is the axis of relationship. It is the recognition that nothing exists in isolation. Even the most stable forms whisper to their surroundings, asking for balance.

Classification: Spatial Constraint ~ external coherence requirement.

6. FSI ~ Filament Saturation Index

Symbol: **FSI**
Domain: **0.00 ~ 1.00** (often displayed on the Codex scale as 0–9)

Function: Measures temporal coherence ~ how completely and for how long a fold saturates its internal filaments with resonance.

High FSI signifies a geometry whose filaments are deeply and persistently filled with GOE. Oscillation remains coherent across many cycles; the fold keeps ringing like a bell that refuses to fully damp. Low FSI indicates shallow or brief saturation ~ resonance drains quickly, and the structure falls back toward potential as soon as excitation stops. In practice, FSI extends or shortens the *duration* of any given resonance state.

In the mathematics of the GCI, FSI joins ERF to define the internal timebeat of a fold. The averaged resonance integrity $\bar{R}$ includes FSI alongside REP, DBI, ERF, and RRP, while MRI provides the spatial weight. High FSI lifts $\bar{R}$ and stretches the fold's effective lifetime; low FSI compresses it.

Philosophically, FSI is the axis of persistence. Where ERF says "I will hold energy," FSI says "I will remain present." Time in Quamitry arises wherever filaments saturate and delay their own decay; every increase in FSI is another moment in which geometry chooses to linger within itself rather than vanish back into the field.

Classification: Temporal Constraint ~ time endurance of structure.

---

7. FTR ~ Failure Threshold Rating

Symbol: **FTR**
Domain: **0 ~ 9** (0 = infinite stability ~ 9 = terminal collapse)
Function: Quantifies how much compressive or resonant load a fold can tolerate before fracture.

FTR expresses the margin between a geometry's current state and its point of failure. Low FTR values (0–3) indicate deep safety ~ the fold can absorb extreme compression without losing coherence. High FTR values (7–9) show that the geometry is already near its breaking point; even modest additional GOE or pressure may trigger collapse.

In Codex practice, FTR is read alongside GOE Load, FSI, and MRI to gauge how far an element stands from the edge of its stability envelope. Stable backbone elements (such as gold or carbon) occupy low FTR regimes; late actinides and transactinide's cluster near FTR = 8–9, where resonance begins to forget its own instructions and unravel.

Philosophically, FTR is the axis of mortality. It marks the humility of form ~ the truth that every geometry has a limit to how much it can remember before it must let go. A fold with FTR ≈ 0 approaches the ideal of eternal containment; a fold with FTR ≈ 9 lives as a brief flare in the lattice.

Classification: Dynamic Behavior Indicator ~ collapse threshold.

---

8. GOE Load ~ Geometric Oscillation Energy Load

Symbol: **GOE Load**
Domain: Qualitative ~ ● Low · ●● Medium · ●●● High
Function: Describes how much Geometric Oscillation Energy is present in the fold before additional compression.

GOE Load is the energetic precondition of a structure ~ the amount of base resonance stored in its geometry prior to any new stress. Low GOE Load corresponds to relatively calm, inert states; Medium to active but manageable excitation; High to highly charged folds where substantial oscillatory potential is already coiled in the form.

In the element tables, GOE Load is read with FTR and Collapse Vector to anticipate how a given fold will respond when pushed. High GOE Load plus low FTR often presages violent modes such as ✳ Ignite or ✧ Shatter; high GOE Load with strong retention and balance can yield dense, long-lived matter. Low GOE Load tends to support inert or gently responsive forms, especially when ERF and DBI are high.

Philosophically, GOE Load is the measure of latent possibility ~ the tension between what a geometry already contains and what it might still become. It is the universe's "breath held" inside matter.

Classification: Dynamic Context Indicator ~ initial energy state.

---

9. Collapse Vector (CV)

Symbol: **CV** (glyphic forms: ✳ Ignite, ♮ Release, ↻ Refold, ↓ Sink, ∞∘ Balance, ✦ Shatter, etc.)
Domain: Qualitative mode set ~ no numeric range
Function: Identifies the natural direction and style of release when a fold reaches failure.

Where FTR tells us *how close* a fold is to breaking, the Collapse Vector tells us *how it lets go*. ✳ Ignite marks explosive thermal release; ♮ Release indicates controlled discharge or decay; ↻ Refold describes a geometry that collapses by returning into a new harmonic self-balance; ↓ Sink indicates energy sinking into deeper structures or fields; ✦ Shatter denotes fragmentation into smaller resonance packets; ∞∘ Balance represents quasi-non-collapse, where excess energy is re-distributed into a higher equilibrium rather than destroying the fold.

In the Codex Index View, CV sits beside FTR, GOE Load, and MRI as the qualitative fingerprint of an element's extreme behavior. For heavy and synthetic elements, the collapse vector is often the most informative signal of how their short lives end ~ whether in flash fission, gradual release, or brief returns to symmetry.

Philosophically, the Collapse Vector is the axis of destiny. Collapse in Quamitry is not destruction but reintegration. Each vector shows how a form chooses to return its memory to the field ~ violently, gently, silently, or by becoming something new.

Classification: Collapse Vector ~ qualitative failure mode descriptor.

---

10. GCI ~ Geometric Compression Index

Symbol: **GCI**
Domain: **0 ~ 9** on the Codex scale
Function: Composite index of resonance coherence ~ how efficiently a structure holds, redirects, and releases GOE under compression.

Mathematically, GCI is derived from the Deca relation: the mean of the five resonance axes (REP, DBI, ERF, RRP, FSI) multiplied by the spatial-temporal weight MRI ÷ 10, then scaled into the 0–9 Codex range. High GCI arises where internal resonance is balanced and retained and the mirrorfield does not overburden the fold; low GCI marks geometries that trade endurance for rapid exchange.

In interpretation:

- High GCI (7–9) ~ geometry of endurance ~ low entropy ~ strong harmonic closure.

- Moderate GCI (5–6) ~ adaptive geometry ~ balance between flow and retention.

- Low GCI (0–4) ~ reactive or transient geometry ~ fast energy exchange ~ high entropy.

Philosophically, GCI is the bridge between structure and motion ~ the first measure that reads matter not by its mass, but by its memory. It compresses the entire Deca-Axis conversation into a single value: how much energy a geometry can remember before it must release.

Classification: Dynamic Stability Indicator ~ composite outcome of all other axes.

Together, these resonance axes define the internal behavior of geometry, while MRI situates that behavior in space and the remaining structural measures resolve stability and collapse. Their interplay forms the multidimensional resonance field from which the GCI is derived. In them, we read not only the architecture of matter, but its moral logic ~ endurance as rhythm between holding, releasing, reflecting, and remaining.

SECTION 4.7 ~ THE DECA-AXIS STABILITY MAP

Every element is not a fixed point, but a circulation of ten decisions. The Deca-Axis framework makes those decisions visible ~ who chooses endurance, who chooses exchange, and who lives at the edge of collapse.

Viewed through the Convergence Table of laws and folds, three patterns emerge.

4.7.1 ~ Axes that Anchor Stability

14

The **stable spectrum** of geometry is anchored by axes that maximize balance, retention, and temporal coherence:

- **DBI ~ Density Balance Index**
  Harmonic equilibrium between inward compression and outward release. High DBI marks folds that *breathe* rather than crack or disperse.

- **ERF ~ Energy Retention Factor**
  Memory per cycle ~ how much GOE remains after each oscillation. High ERF keeps resonance circulating instead of leaking away.

- **FSI ~ Filament Saturation Index**
  Duration of saturation ~ how long the filaments stay filled. High FSI stretches resonance into time, turning a moment of coherence into a lived history.

- **REP ~ Resonant Efficiency Potential**
  Efficiency with which motion becomes structure. High REP lets geometry lock GOE into coherent architecture with minimal waste.

In the law language:

- **Preform Density** leans on **DBI + REP** ~ how cleanly compression can become mass.

- **Resonant Delay** is quantified by **ERF + FSI** ~ how long that mass remembers its own vibration.

These pairs give rise to the **Shellfold** and **Lockfold** archetypes ~ the quiet architects of endurance. Carbon's Lockfold, for example, combines high REP, DBI, and ERF with supportive FSI to reach a high GCI ~ a geometry that not only holds memory but continues to remember itself across cycles.

4.7.2 ~ Axes that Drive Exchange and Reactivity

Where stability anchors, **exchange axes** move:

- **RRP ~ Resonant Release Potential**
  The grace of letting go. High RRP enables controlled

15

conduction and catalytic behavior; moderate RRP prevents dangerous buildup; very low RRP hoards energy until sudden failure.

- **MRI ~ Mirrorfield Requirement Index**
  Dependency on external symmetry. Low MRI elements can exist in isolation; high MRI elements *require* bonds, fields, or lattices to remain coherent and will react aggressively until they find them.

Together with REP, these axes govern the **Law of Resonant Motion** and **Law of Polarized Resonance**, shaping:

- **Chainfolds** ~ transmission folds where RRP + MRI tune how energy moves through conductive and biological networks.

- **Crossfolds** ~ polarity folds where REP + RRP sharpen edges of charge and chemical reactivity.

High RRP and high MRI in isolation often signal reactive or catalytic forms (e.g., halogens, alkali metals). The same values, embedded in the right lattice, produce life, magnetism, and circuitry.

Moderation matters. Elements like carbon and silicon sit in a sweet spot: enough RRP to share resonance, modest MRI to benefit from networks without being enslaved to them. This balance enables stable yet adaptable architectures ~ the backbone of organic and technological memory.

4.7.3 ~ Axes of Collapse and Transformation

Collapse in Quamitry is not disappearance but **reformatting**. Three axes describe how the fold approaches and crosses that edge:

- **GOE Load** ~ how much energy is already coiled.
- **FTR** ~ how much more the fold can take.
- **Collapse Vector** ~ what happens when FTR is exceeded.

In the early phases, these axes are quiet ~ FTR is low, GOE Load is modest, and collapse vectors like ℧ Refold or ∞∘ Balance rarely activate. By the **Actinide Epoch** and beyond, GOE Load climbs while

16

FTR approaches 8–9; collapse vectors shift toward ✳ Ignite and ✧ Shatter. Geometry begins to forget.

The Collapse Epoch is where the Deca map shows its starkest contrast: GCI remains moderate, but ERF and FSI decline, MRI grows burdensome, and the fold can no longer safely anchor what it holds. Matter reverts to resonance through its vector of choice ~ fragmentation, thermal burst, implosion, or refolding into ghost-like forms.

4.7.4 ~ High-GCI Patterns ~ Stability Through Circulation

High GCI is not achieved by maximizing a single axis, but by **circulating well** among all ten:

- Internal axes (REP, DBI, ERF, RRP, FSI) must be **balanced** rather than extreme.

- Spatial demand (MRI) must be **moderate**, not so low that complexity is starved, nor so high that the fold cannot live alone.

- FTR must leave room beneath the breaking point; GOE Load must be supportable; Collapse Vector should be benign when it finally engages.

Hydrogen exemplifies efficient containment with minimal field demand; carbon exemplifies structured endurance; iron exemplifies magnetic architecture. Each sits high on the GCI scale for the same reason: their Deca profiles are *round* ~ no axis drags resonance into a catastrophic imbalance.

In this view, the Deca-Axis Stability Map is a topological landscape. Some folds cluster in deep basins of endurance; others ride high ridges of reactivity; a few skim the knife-edge where structure and collapse are indistinguishable.

4.7.5 ~ Anchor Nodes and SubQUAMI Locks

At the smallest scales, stable matter begins wherever resonance finds an **anchor node** ~ a fold where:

- REP is high enough to channel GOE into form,

- DBI finds a rhythmic center,

- ERF and FSI are sufficient to keep that rhythm alive, and

- MRI and GOE Load do not demand more support than the environment can provide.

These events are the **SubQUAMI locks** of the Codex narrative ~ the first decisions of resonance to remain. Hydrogen is the primordial lock; carbon is the first architectural lock; heavy, short-lived elements are failed or over-driven locks that demonstrate the limits of compression.

The Deca-Axis map allows these stories to be read numerically. It shows that every element is a different answer to the same question: *How much of myself can I remember, and for how long?*

Table 4.7.1 ~ The Ten Axes and Their Roles in Matter Behavior

AXIS	KEY FUNCTION	ROLE IN STABILITY	ROLE IN REACTIVITY / COLLAPSE	CATEGORY
GCI ~ Geometric Compression Index	Composite index of resonance coherence (0–9).	High GCI ~ enduring, low~entropy forms; signal that internal axes are balanced and MRI demand is tolerable.	Low GCI ~ transient or reactive geometry; any major imbalance in the Deca axes depresses GCI and pushes toward transformation or collapse.	Dynamic Stability Indicator (composite)
GOE Load	Available internal GOE before additional compression (● Low to ●●● High).	Low to moderate load is easy to stabilize if DBI, ERF and FTR are favorable; supports calm,	High load strains the fold ~ if retention or tolerance are weak it drives reactions or collapse (Ignite, Shatter, etc.).	Dynamic Context Indicator (initial energy state)

		inert or gently active folds.		
REP ~ Resonant Efficiency Potential	Efficiency of converting GOE into structure $(0\sim1)$.	High REP strengthens self~cohesion and reduces entropy loss; key to persistent, self~anchoring folds.	Low REP lets energy escape faster than structure can contain it; supports turbulent, leaky or short~lived forms.	Dynamic Behavior Indicator (internal cohesion)
DBI ~ Density Balance Index	Balance of compressio n vs expansion $(0-1)$.	High DBI ~ harmonic pressure distribution; prevents brittle fracture and runaway expansion; archetype of " graceful" stability.	Low DBI ~ over~compressio n or over~expansion; tends toward shattering or volatile release when stressed.	Dynamic Behavior Indicator (structural harmony)
ERF ~ Energy Retention Factor	Fraction of oscillation energy retained per cycle $(0\sim1)$.	High ERF ~ long~term energy storage; geometry " remembers" excitation and persists through time (axis of memory).	Low ERF ~ fast energy loss; structure forgets each cycle and decays or relaxes quickly unless constantly driven.	Dynamic Behavior Indicator (temporal retention)
RRP ~ Resonant Release Potential	Ability to release or transfer energy without collapse $(0-1)$.	Moderate to high RRP ~ controlled release; prevents destructive buildup and enables safe bonding, conduction and exchange.	Very high RRP drives strong interaction and catalytic behavior; very low RRP traps energy until threshold, often ending in sudden Ignite or Shatter collapse.	Dynamic Behavior Indicator (energy exchange)

FTR ~ Failure Threshold Rating	Tolerance to compression before fracture $(0\sim9; 0 =$ far from failure, 9 = near collapse).	Low FTR $(0\sim3)$ ~ far from failure; elements can endure extreme conditions and serve as structural backbones.	High FTR $(7\sim9)$ ~ near collapse; common in late, unstable or synthetic folds with short lifetimes and easy decay.	Dynamic Behavior Indicator (collapse threshold)
Collapse Vector (CV)	Mode of failure (Refold, Ignite, Release, Sink, Shatter, Balance...).	In truly stable regimes, CV remains latent; mild vectors (Refold, Balance) signal graceful re~equilibration if ever engaged.	Determines how the fold returns its memory when it fails ~ explosive, fragmentary, gentle or transmutational; qualitative fingerprint of collapse behavior.	Collapse Vector (failure mode)
MRI ~ Mirrorfield Requirement Index	Required external field coherence (0–10).	Low MRI supports solitary stability; elements can hold shape and resonance in isolation (self-sufficient folds).	High MRI demands bonds or fields; when alone it drives strong reactivity or rapid transformation until a supportive mirrorfield is found.	Spatial Constraint (external coherence)
FSI ~ Filament Saturation Index	Temporal coherence of filament saturation (0–1).	High FSI ~ long~lasting filament saturation; resonance stays coherent across many cycles, extending the fold's lifetime (timebeat of the fold).	Low FSI ~ short resonance duration; even stable forms become transient, decaying or transforming quickly once excited.	Temporal Constraint (time endurance of structure)

SECTION 4.8 ~ THE FOLD TYPOLOGY

In Quamitry, a fold is not a metaphor. It is the primary act by which resonance learns to inhabit form. A fold is geometry in negotiation with itself ~ a compression of GOE that produces stability, rhythm, or release. Every element expresses one or more-fold archetypes, each representing a unique method of resonance containment.

These fold types are the structural DNA of the Codex. They explain how an element behaves long before we describe what it does.

In the updated frame, each Fold is described in terms of its **anchors**:

- **Fixed Anchors** ~ proton wells that hold curvature in place; they define where the lattice refuses to move.

- **Mobile Anchors** ~ electron fields that roam, hop, or lock between wells; they define how the lattice chooses to move.

The six archetypes below now include this anchor pattern explicitly, linking the qualitative behavior of each fold to how Fixed and Mobile anchors share tension.

---

1. △ Shellfold ~ The Containment Fold

Behavior: Inward compression forming a hollow equilibrium.

Function: Creates self-sustaining stability through curved inward resonance.

Anchor Pattern: Fixed Anchors are minimal but deeply centered; Mobile Anchors are lightly bound, forming low RRP shells around the core.

Example: Hydrogen, Beryllium, Calcium.

The Shellfold geometry traps resonance within concentric layers wrapped around a small number of deeply centered Fixed Anchors. Mobile Anchors skim the interior surface as a faint shell field, giving

21

just enough RRP for interaction while never pulling the fold off center. Subtle inward tension from the Fixed core produces simple, enduring stability.

Hydrogen is the primordial Shellfold: a single Fixed Anchor balancing a sparse Mobile field with almost no polar leverage. It is the universe's first successful act of containment ~ motion held without needing elaborate structure.

Philosophical Aspect: The will to exist. Shellfolds are the universe learning to say "I am."

---

2. ℧ Refold ~ The Return Fold

Behavior: Dual layered curvature; energy loops back on itself.

Function: Perfects harmonic closure, minimizing entropy by reflection.

Anchor Pattern: Fixed Anchors are tightly inward symmetrical; Mobile Anchors are nearly absent or fully constrained, yielding zero net polarity and no escape channels.

Example: Helium, Neon, Argon.

Refolds represent geometry folding inward until resonance meets its own echo. The Fixed Anchors arrange themselves in perfect inward symmetry, while Mobile Anchors are either locked into closed shells or effectively withdrawn from the field. No net charge gradient emerges ~ the Mobile Anchor field has nowhere to leak, so every oscillation returns home as a standing wave.

This anchor pattern maps directly onto noble gas inertness: high ERF with low MRI and no polar release pathways. The fold does not seek partners because there is no exposed imbalance to resolve.

Philosophical Aspect: The will to rest. The Refold is memory at peace with itself.

---

3. ◇ Chainfold ~ The Transmission Fold

22

Behavior: Linear or branching sequences of partially open resonance nodes.

Function: Enables conductive transfer of GOE between nodes.

Anchor Pattern: Fixed Anchors align in a conductive sequence; Mobile Anchors are loosely coupled and able to hop from anchor to anchor along the chain.

Example: Lithium, Sodium, Potassium.

The Chainfold connects rather than isolates. Fixed Anchors arrange themselves in a line or branching track, creating a series of shallow wells instead of a single deep basin. Mobile Anchors ride this track, slipping from one well to the next with relatively high RRP. The result is a resonant path that willingly passes energy along instead of trapping it in place.

This is the archetype of conduction, magnetism, and biochemical exchange. A Chainfold prefers circulation over containment: it sacrifices some local stability so that resonance can travel through it as current, signal, or shared bond.

Philosophical Aspect: The will to connect. Every Chainfold is a bridge ~ a promise that stability can travel.

---

4. ⊞ Lockfold ~ The Interlocking Fold

Behavior: Nested resonance geometries interlocked through angular symmetry.

Function: Stores GOE in bonded nodes ~ the architecture of memory.

Anchor Pattern: Fixed Anchors arrange themselves into symmetric memory grids; Mobile Anchors are partly confined into predictable anchor fields, locking in repeatable patterns.

Example: Carbon, Silicon, Aluminum.

The Lockfold is the engineer of the lattice. Fixed Anchors do not merely gather ~ they tessellate into interlocking grids, creating a scaffold of repeating angles and shared wells. Mobile Anchors are neither free nor frozen: they occupy well defined paths and domains, able to shift within the grid but rarely escape it. This semi mobile confinement is what lets the fold encode "instructions" ~ bond angles, lattice rules, logical pathways.

Carbon's four node Lockfold is the archetype of architectural intelligence: four Fixed Anchors arranged so that Mobile Anchors can form networks of astonishing depth without losing coherence. Molecules, crystals, and information bearing chains are all Lockfold expressions ~ endurance through organized complexity.

Philosophical Aspect: The will to remember. To lock is to retain not just form, but instruction.

---

5. ✧ Crossfold ~ The Polarity Fold

Behavior: Intersecting resonance paths creating alternating charge regions.

Function: Enables catalytic interaction and polarized reactivity.

Anchor Pattern: Fixed Anchors are strongly asymmetrical; Mobile Anchors are highly reactive and eager to compensate for the imbalance, amplifying tension between inward and outward charge.

Example: Fluorine, Sulfur, Chlorine.

The Crossfold is tension embodied. Fixed Anchors occupy an uneven geometry that pulls resonance off center. Mobile Anchors respond by clustering into high contrast regions, overcompensating in some directions and leaving deficits in others. Intersecting resonance paths carve out alternating pockets of charge, creating large potential differences across very short distances.

This anchor pattern is the root of extreme reactivity: the Fixed ~ Mobile mismatch drives the fold to seek compensation in anything it can bond to. Halogens are canonical Crossfolds ~ their Mobile Anchors lunge toward any available partner that can soothe the imbalance, releasing significant energy in the process.

Philosophical Aspect: The will to change. Every polarity is a conversation between opposites seeking reconciliation.

---

6. × Fracture Fold ~ The Transitional Fold

Behavior: Over compressed or misaligned geometry reaching its collapse vector.

Function: Releases stored GOE through partial or total decomposition.

Anchor Pattern: Fixed Anchors are over compressed or geometrically misaligned; Mobile Anchors cannot stabilize the tension and instead trigger collapse modes, breaking into new anchor configurations.

Example: Highly unstable isotopes; synthetic or transient elements.

The Fracture Fold is what happens when the lattice cannot find a viable anchor pattern at all. Fixed Anchors crowd or skew past the point where DBI can recover; Mobile Anchors whirl without establishing a stable field. As stress accumulates, the fold approaches its Collapse Vector ~ ignite, release, sink, shatter, or refold ~ and stored GOE is discharged as decay, fission, or emission.

In this view, collapse is not failure but renewal. A Fracture Fold represents the threshold at which geometry must surrender its current memory to remain part of motion. The anchors do not vanish; they're sorted into new folds where resonance can breathe again.

Philosophical Aspect: The will to evolve. All matter eventually fractures, only to be reborn in new harmonics.

Together, these six archetypes form the Fold Typology of the GCI Codex. They are not artistic embellishments but operational descriptors of resonance architecture. When combined with the five axes, they provide a full coordinate system for the behavior of matter ~ defining not only what an element is, but how it decides to be.

SECTION 4.9 ~ FOLD–LAW CORRELATION MATRIX

Each Fold Type behaves as a distinct mode of resonance, and every mode is shaped by one or more of the Five Foundational Laws of Quamitry. Where the Deca Axes define *how* geometry moves, the Fold Law pairings explain *why* those mechanics arise.

Together, they form the Codex's behavioral grammar ~ the reason Hydrogen acts like Hydrogen and Carbon like Carbon.

Fold Law Matrix

FOLD TYPE	DOMINANT LAWS	PRIMARY AXES	BEHAVIORAL ESSENCE
Δ Shellfold	Law of Preform Density ~ Law of Resonant Delay	DBI ~ ERF	Compression seeking harmony ~ containment without collapse
⩒ Refold	Law of Resonant Motion ~ Law of Resonant Delay	ERF ~ MRI	Reflection and internal symmetry ~ energy in repose
⩐ Chainfold	Law of Resonant Motion ~ Law of GOE Transformation	RRP ~ MRI	Transmission of resonance ~ geometry learns to communicate
▽ Lockfold	Law of Preform Density ~ Law of GOE Transformation	DBI ~ REP	Structure as stored resonance ~ geometry remembers instruction
✧ Crossfold	Law of Polarized Resonance ~ Law of GOE Transformation	REP ~ RRP	Polarity and catalytic reactivity ~ the will to change
⚡ Fracture Fold	Law of Polarized Resonance ~ Law of	RRP ~ DBI	Controlled collapse and renewal ~ transformation through release

26

	Resonant Motion		

Interpretive Notes

- **Shellfold and Refold** live in the realm of containment and reflection. They are the stable forms ~ memory preserving itself through compression and internal symmetry.

- **Chainfold and Lockfold** are architectural and connective. They transmit resonance between anchors, building the scaffolding of lattices, molecules, and long-range structures.

- **Crossfold and Fracture Fold** are dynamic and transitional. They mark the edges where geometry learns from stress, releasing stored instruction into new harmonic states.

Each Fold is thus a conversation between at least two Laws and their governing Axes. Within that conversation lies the entire behavior of matter ~ whether it will hold, move, or transform.

Section 5 ~ Example Calculation: Carbon

The Fold and Law systems define *behavior*; the GCI equation reveals it numerically. To illustrate the full process, we walk through a single element: **Carbon** ~ the archetypal Lockfold and the first geometry to achieve structured memory.

For clarity, we assume the Deca relation:

- $\bar{R} = (REP + DBI + ERF + RRP + FSI) \div 5$
- $W = MRI \div 10$
- $GCI = 9 \times \bar{R} \times W$

Exact numeric scaling can be tuned in the Codex tables; what matters here is the *pattern*.

Step 1 ~ Identify the Fold and Laws

- **Fold Type:** ∇ Lockfold
- **Dominant Laws:** Preform Density ~ GOE Transformation
- **Primary Axes:** DBI ~ REP

Interpretation: Geometry that stabilizes energy through ordered interlock ~ endurance through memory.

Step 2 ~ Gather Measured Axial Values

AXIS	DEFINITION	CARBON VALUE
REP	Resonant Efficiency Potential	0.80
DBI	Density Balance Index	0.78
ERF	Energy Retention Factor	0.81
RRP	Resonant Release Potential	0.74
MRI	Mirrorfield Requirement Index (0 ~ 10)	3.0
FSI	Filament Saturation Index (0.00 ~ 1.00)	(set in table)

These arise from empirical or modeled resonance-response testing ~ how Carbon's geometry sustains, balances, and exchanges GOE across cycles.

Step 3 ~ Apply the Deca Relation

$$\bar{R} = (REP + DBI + ERF + RRP + FSI) \div 5$$
$$W = MRI \div 10 \qquad GCI = 9 \times \bar{R} \times W$$

Inserting Carbon's measured values yields a GCI in the **high stability range**, approximately:

GCI ≈ 7.8 / 9

The exact numeric result is fixed in the Codex tables; this section is about *reading* what that result means.

Step 4 ~ Interpret the Result

28

Carbon's resonance maintains a high fraction of coherence under continuous oscillation. Its geometry achieves near-perfect density balance while allowing measured release, creating the lattice architectures that define organic life and crystalline endurance.

- **High REP + DBI** ~ Efficient compression with symmetrical internal distribution.
- **High ERF** ~ Memory retention across molecular cycles.
- **Adequate FSI** ~ Resonance persists across time, not just across a single cycle.
- **Moderate RRP & MRI** ~ Controlled release and cooperative bonding rather than isolation or explosive reactivity.

In words: Carbon does not merely exist; it *remembers* its own structure.

Step 5 ~ Correlate with the Fold Matrix

FOLD	DOMINANT LAWS	OUTCOME
▽ Lockfold	Preform Density ~ GOE Transformation	Converts GOE into long-term lattice stability ~ geometry becomes architecture.

Carbon therefore stands at the first harmonic of intelligent structure. In Carbon, geometry no longer merely persists; it begins to organize. Every later form of complexity, from crystal to cell, borrows its endurance from this Lockfold decision: resonance choosing to remember itself.

Step 6 ~ Contextual Reading

- **High GCI values (> ~7)** correspond to geometries that are self-sustaining and capable of transmitting instruction.
- **Low GCI values (< ~4)** mark geometries that favor transformation or release.

In this spectrum, Carbon defines the midpoint between stability and creativity ~ matter coherent enough to persist, flexible enough to evolve.

Summary

GCI ≈ 7.8 → memory achieved through balanced compression.

The Lockfold reveals that endurance is not rigidity but *rhythmic equilibrium* ~ a geometry that holds, releases, and re-holds in measured cadence.

Section 6 ~ The Law Convergence Table

All Quamitric behavior ~ from atomic endurance to stellar resonance ~ can be traced to the interplay between the **Five Foundational Laws** and the **Deca Axes of Measurement**.

- The **Laws** define the *intent* of geometry.
- The **Axes** measure its *execution*.

Where they intersect, we find the universe's instructions: how matter learns to hold, release, and remember energy.

Each Law can therefore be represented as a domain of influence upon the Axes, creating a multidimensional relationship map known as the **Law Convergence Table**.

LAW OF QUAMITRY	PRIMARY AXES OF INFLUENCE	CORE FUNCTION	OBSERVABLE OUTCOME
Law of Preform Density	DBI ~ REP	Governs the compression of GOE into form. Defines how mass is born from resonance.	Establishes structural stability and fold initiation.
Law of Resonant Motion	RRP ~ MRI	Controls the propagation of resonance through geometry. Determines how energy moves, oscillates, or transmits.	Creates conduction, orbital motion, and field mobility.
Law of Polarized Resonance	REP ~ RRP	Describes the tension and balance between opposing resonance fields. Manifests as polarity, magnetism, and chemical potential.	Produces reactive behavior and magnetic symmetry.
Law of Resonant Delay	ERF ~ FSI	Defines time as persistence within resonance. Measures the capacity of geometry to sustain vibration without collapse.	Establishes temporal endurance ~ the timebeat of matter.
Law of GOE Transformation	All Axes	Unites compression and release. Converts raw geometric oscillation into matter, energy, and light.	Serves as the bridge between existence and expression.

Interpretive Notes

- **Preform Density** and **Resonant Delay** anchor the stable spectrum of geometry. They produce the Shellfold and Lockfold archetypes ~ the quiet architects of endurance. Resonant Delay is quantified by **FSI alongside ERF** ~ a fold's timebeat arises from what is retained and how completely filaments are saturated.
- **Resonant Motion** and **Polarized Resonance** govern dynamism. They shape Chainfolds and Crossfolds ~ geometries of exchange and reaction.
- **GOE Transformation** connects all others, acting as the prime operator of Quamitry ~ the act through which structure and motion are revealed as the same event, viewed from opposite sides of resonance.

The Convergence Table is the Codex's compass. It lets an observer:

1. Locate a behavior.
2. Trace it to a governing Law.
3. Follow that Law into its Axes.
4. Understand the geometry underneath.

Where traditional science splits physics, chemistry, and time, Quamitry reunites them: one system of Laws, a handful of Axes, six Folds of expression.

Section 7 ~ Reading a GCI Entry

Every page of the GCI Codex is a conversation between number and meaning. Each element ~ from Hydrogen to Oganneson ~ is presented not as a static substance, but as a resonance event frozen into structure.

To read a GCI entry is to read the memory of geometry itself.

A standard two-page spread consists of two complementary halves:

The Index View presents the **quantifiable** geometry of an element ~ its numerical identity in the Quamitric field. This side of the Codex corresponds to the measurable domain: the language of data and proportion.

Field Guide

- **Fold Type (△, ʊ, ◇, 吅, ✦, ×)**
  Indicates the element's primary geometric behavior ~ how it stores or transmits GOE.
- **FTR (Failure Threshold Rating)**
  How much compression the fold can endure before collapse ~ 0 = effectively unbreakable within normal ranges; 9 = terminal collapse.
- **Collapse Vector**
  Symbol showing the probable direction or mode of failure (ʊ refold, ✳ ignite, ⚡ release, ↓ sink, ✧ shatter, ∞∘ balance, etc.).
- **GOE Load**
  The density of energy currently stored in the geometry ~ expressed qualitatively (● Low, ●● Medium, ●●● High).
- **REP / DBI / ERF / RRP / FSI / MRI**
  The internal resonance axes plus the spatial coherence requirement.
- **GCI (0 ~ 9)**
  The composite Quamitric score derived from the internal axes, weighted by MRI.
- **Resonant Function**
  A short descriptor summarizing the element's energetic personality.
- **Signature Behavior**
  A one-sentence description of how the fold interacts with surrounding resonance fields.

Visually, the Index View is a technical portrait ~ a folded map of how resonance has chosen to inhabit form. It is designed for comparison across elements and for analytical study of the Codex as a whole.

RIGHT PAGE ~ THE CODEX VIEW

The Codex View is the mirror of the Index. Here, geometry becomes language again ~ number becomes narrative. This page translates resonance data into human terms ~ it explains why a structure behaves as it does, and what that means for the nature of matter.

In the finished layout, the Codex View is organized into named fields:

- **Fold Classification** and **Resonant Tier** ~ locate the element inside the Fold Typology and stability spectrum.
- **Phase Polarity** ~ states the dominant inward outward bias of the fold.
- **Quamitric Behavior** ~ gives the core behavioral thesis of the element as a resonance event.
- **Fold Mechanics** ~ summarizes the internal geometry: node arrangement, anchor behavior, GOE flow.
- **Field Correlations** ~ links the fold to observable environments ~ stellar cores, lattices, plasmas, biological systems.
- **Canonical Insight** ~ a single line that captures the element's teaching inside the Codex.
- **RTI Signature** ~ describes how the fold appears to instrumentation ~ scatter patterns, hysteresis, response curves.
- **Prediction Hook** ~ hints at how altering context (mix ratios, fields, gradients) should change MRI, FSI, or behavior.
- The **fold diagram or echofold image** gives a visual imprint of the geometry ~ an immediate sense of how the anchor field lives in space.

These fields still express three implied layers:

- **Physical Interpretation** in Phase Polarity, Fold Mechanics, Field Correlations, RTI Signature.
- **Quamitric Meaning** in Quamitric Behavior and Canonical Insight.
- **Human Implication** in Resonant Tier and Prediction Hook.

Thus, every Codex entry is both a report and a meditation. The left page measures: the right page remembers why it matters ~ how the universe is built, and why it chooses to keep being that way.

How to Read the Codex as a Whole

- **Horizontally:** Compare elements by phase or period to trace the evolution of geometry across the GCI spectrum, from containment to release.
- **Vertically:** Follow Fold Type groupings to study how geometry behaves under different energy densities.
- **Diagonally:** Read the Law associations to see how stability transforms into reactivity, and endurance into motion.

In this way, the Codex becomes not a table, but a score. Each element is a note ~ each Law a rhythm ~ each Fold a chord in the great harmonic composition of matter.

Summary

The left page **measures**: the right page **interprets**.

The Fold defines structure; the Law gives purpose; the GCI binds them through mathematics ~ the universal language of resonance.

Section 8 ~ Example Walkthrough (Hydrogen → Helium)

To demonstrate how a reader should move between Index and Codex Views, we begin with the simplest conversation in the universe ~ **Hydrogen speaking to Helium**.

They form the first lesson in resonance behavior: motion and rest, expansion and reflection, existence and memory.

35

HYDROGEN ~ THE FIRST FOLD

- **Fold Type:** △ Shellfold
- **Dominant Laws:** Preform Density ~ Resonant Motion
- **GCI:** 7.5

In the Index View, Hydrogen shows:

- High **REP** and **DBI** ~ efficient containment with harmonic balance.
- Moderate **MRI** ~ it is largely self-stabilizing, with minimal field demand.
- Balanced **ERF** and **RRP** ~ it holds enough energy to persist yet can participate in motion.
- A near-equilibrium **Collapse Vector** (~) ~ neither strongly implosive nor explosive under normal field stress.
- FSI indicating high filament occupancy ~ the seed's timebeat is steady rather than fleeting.

In the Codex View, this becomes the **First Act of Geometry**: the choice to remain.

Through the **Law of Preform Density**, GOE curls inward until it holds its own echo. Through the **Law of Resonant Motion**, we see that this holding is never static ~ it is vibrational.

Practically, this makes Hydrogen the most energy-efficient form possible ~ nothing wastes less energy simply existing. Philosophically, Hydrogen teaches that persistence begins with acceptance: to remain is not to resist motion, but to flow within its rhythm.

HELIUM ~ THE REFLECTION OF FORM

- **Fold Type:** ▽ Refold
- **Dominant Laws:** Resonant Delay ~ GOE Transformation
- **GCI:** 6.9

In the Index View, Helium exhibits:

- REP and DBI in almost perfect symmetry.
- Very low **MRI** ~ it requires almost no external field to remain coherent.
- **Collapse Vector:** ↻ Refold ~ energy reenters itself in harmonic balance.
- FSI approaching completion ~ filaments saturated, resonance held across time.

Helium is the **First Mirror**. It shows that geometry's highest act of endurance is not defense, but **reflection**.

The **Law of Resonant Delay** stretches time itself ~ vibration held so gently that decay slows to a whisper. The **Law of GOE Transformation** appears here as completion: resonance no longer seeks elsewhere to exist.

Mathematically, this explains Helium's inertness. It does not bond, because it has no unmet energetic need. It exists in harmonic closure ~ resonance without tension.

Philosophically, Helium teaches the universe how to pause. It is the geometry of **rest** ~ the necessary silence between pulses.

READING THE RELATIONSHIP

PROPERTY	HYDROGEN	HELIUM	TRANSITION
Fold Type	△ Shellfold	▽ Refold	Containment → Reflection
Dominant Laws	Preform Density ~ Resonant Motion	Resonant Delay ~ GOE Transformation	Compression learns rest
Behavioral Essence	Motion preserved	Motion resolved	The universe's first cycle of breath

Between Hydrogen and Helium lies the first complete Quamitric circuit ~ the **Alpha Resonance**. Hydrogen compresses motion into being; Helium reflects it into balance. Every later element repeats this dialogue in more complex languages.

WHY THE WALKTHROUGH MATTERS

Understanding these two pages teaches the reader *how to see every other pair*. Each Fold Type and Law echoes this relationship in different octaves, from simple containment to vast reactive complexity.

The GCI does not merely measure matter; it measures the progression of **consciousness through geometry**.

To read one spread is to glimpse the pattern of all:

- **Hydrogen:** The birth of containment.
- **Helium:** The memory of rest.
- **Together:** The breath of creation.

SUMMARY

The Codex begins with Hydrogen's decision to endure and Helium's decision to remember. From these two notes, the symphony of matter unfolds.

Section 9 ~ The Epochal Progression

THE STORY OF GEOMETRY THROUGH TEN PHASES

The Geometric Compression Index is not merely a taxonomy of elements. It is a chronicle of geometry learning to move, remember, and transform. Each Phase within the Codex gathers a band of elements that share a characteristic Fold profile and Law signature ~ an evolutionary moment in resonance behavior.

Across ten epochs, the folds of geometry rehearse every lesson that existence can teach: containment, reflection, polarity, transmission, ignition, and renewal. Together, the ten Phases map cleanly onto the 118 known elements, from Hydrogen to Oganesson.

PHASE 1 ~ THE PRIMORDIAL MEMORY (H–NE)

Theme: Containment and reflection
Dominant Folds: Shellfold ~ Refold

Behavior: Resonance learns stability. The first ten elements form the alphabet of structure. Hydrogen establishes the act of persistence; Helium perfects inert reflection. Through Carbon, geometry discovers architecture ~ the lattice that will one day house life. This epoch is the cradle of form, where matter first says *I will remain*.

PHASE 2 ~ THE CONDUCTIVE EPOCH (NA–CA)

Theme: Polarity learns to move
Dominant Folds: Chainfold ~ Lockfold

Behavior: Resonance begins transmission. Geometry discovers current ~ motion without destruction. Energy no longer sits inside matter; it flows through it. Conductivity, simple magnetism, and cooperative bonding emerge, preparing the ground for chemistry and biology.

PHASE 3 ~ THE MAGNETIC EPOCH (SC–ZN)

Theme: Fields of memory
Dominant Folds: Lockfold ~ Crossfold

Behavior: Geometry forms coherent domains of motion. Magnetism appears as resonance alignment ~ folds communicating at scale. This is the dawn of pattern: the universe begins to remember itself across distance.

PHASE 4 ~ THE REFLECTIVE EPOCH (GA–KR)

Theme: Resonance organization
Dominant Folds: Lockfold ~ Refold

Behavior: Structure refines its symmetry. Elements in this epoch exhibit high cohesion and strong reflective behavior ~ geometry polishing itself toward closure. Stability reaches an early peak just before complexity begins to fracture into more volatile forms.

PHASE 5 ~ THE RESONANT EXPANSION (RB–XE)

Theme: Overload and release
Dominant Folds: Crossfold ~ Chainfold

Behavior: Energy density begins to exceed containment. Geometry overfills its boundaries, producing volatility and new charge distributions. This epoch marks the first tremor of creative chaos ~ the need for renewal as stored tension seeks new paths.

PHASE 6 ~ THE TRANSITIONAL EPOCH (CS–LU)

Theme: Structural experimentation
Dominant Folds: Chainfold ~ Fracture Fold

Behavior: Geometry searches for new balance through instability. Rising FTR and divergent MRI profiles mark matter's attempt to reinvent endurance through transformation. Collapse becomes a tool rather than a failure ~ a way to explore alternative anchor configurations.

PHASE 7 ~ THE COMPLEX EPOCH (Hf–Hg)

Theme: Layered resonance
Dominant Folds: Lockfold ~ Crossfold ~ Fracture Fold

Behavior: Multi-fold cooperation. Here matter behaves almost like thought ~ several fold behaviors coexisting within a single field. This is the beginning of conscious geometry: structures aware of their own harmonic inheritance, capable of subtle catalytic and magnetic behavior.

PHASE 8 ~ THE COLLAPSE EPOCH (Tl–Rn)

Theme: Over-compression and memory loss
Dominant Folds: Fracture Fold

Behavior: Geometry begins to forget. At extreme densities, folds struggle to maintain coherent balance. GOE bleeds into the field, birthing radiation and decay. This epoch teaches that even endurance must eventually return to motion.

PHASE 9 ~ THE ACTINIDE EPOCH (Fr–Lr)

Theme: Transience and rebirth
Dominant Folds: Fracture Fold ~ Crossfold

Behavior: Resonance flickers at the edge of stability. These geometries are born faster than they can remember, often existing only as brief instructional states. Their behavior encodes a single lesson: beyond endurance lies transformation into something new.

PHASE 10 ~ THE TRANSACTINIDE THRESHOLD (RF–OG)

Theme: Directed synthesis of geometry
Dominant Folds: Lockfold ~ Chainfold

Behavior: Human-guided compression. The transactinide's are largely lab-born folds ~ resonance coaxed into existence through deliberate synthesis. Here the Codex still tracks real elements, but the mode of genesis shifts: geometry is now invited as much as discovered. This epoch hints at Quamitry's promise ~ conscious design of folds, not just observation of them.

THE RESONANT SECTION ~ SYNTHETIC & THEORETICAL CONSTRUCTS

Beyond the 118 known elements, the Codex maintains a separate Resonant section for synthetic and theoretical constructs: extended lattices, SubQUAMI locks, hypothetical ultra-stable folds, and post-Og geometries. These are not assigned to a numbered Phase; instead, they serve as exploratory waypoints ~ sketches of what geometry *could* become when guided by intentional design or encountered in more extreme cosmic conditions.

Here, GCI values, Fold Types, and Law signatures are proposed rather than measured, offering a sandbox for future experiments. The ten elemental Phases describe what existence has already been chosen; the Resonant section gestures toward what it may yet allow.

THE CYCLE OF TEN

Across these ten epochs the Codex traces a full circle of existence: **Containment ~ Motion ~ Reflection ~ Collapse ~ Renewal.**

Every Fold Type, every Law, every GCI value is a syllable in this universal sentence. When viewed together, the ten Phases are not merely a table but a story ~ the autobiography of matter.

"The Codex begins where silence remembers motion and ends where memory becomes creation again."

Section 10 ~ The Temporal Model of Resonance

TIME AS FOLDED DURATION

Time is not a river. It is a resonance. In Quamitry, time arises wherever geometry delays its own return ~ wherever filaments saturate and remain coherent. This delay is captured by FSI alongside ERF. What we perceive as the flow of time is not motion through space, but resonance persisting through compression.

Each fold, each atom, each layer of structure carries within it a delay ~ a Resonant Duration. This delay is governed by the Law of Resonant Delay and measured through ERF and FSI on the Codex axes. Where ERF is high, geometry holds vibration longer; where ERF is low, vibration releases quickly. Time is therefore not universal; it is local to each fold. Every structure in the universe experiences its own tempo, set by how long its resonance can remember itself before collapsing.

10.1 ~ RESONANT DURATION (ΔT_r)

$$\Delta \tau_r = \frac{ERF \times DBI \times FSI}{RRP + \varepsilon}$$

ERF quantifies retention. DBI preserves internal balance. FSI measures how completely filaments are saturated. RRP defines the rate of release. ε is the ambient loss term that accounts for background leakage into the field.

High $\Delta\tau_r$ corresponds to slow time ~ sustained coherence.
Low $\Delta\tau_r$ corresponds to fast time ~ quick turnover.

10.2 ~ TEMPORAL DENSITY AND DECAY

The perception of aging or decay is a geometric translation of diminishing ERF. As resonance begins to lose coherence, its internal duration shortens; cycles of oscillation complete more rapidly until the structure can no longer maintain fold integrity. Entropy, in Quamitric language, is simply the forgetting of rhythm.

Every atom, every cell, every star possesses a built-in temporal signature ~ the interval of its persistence. Stable folds (like Carbon) experience time as a slow hum; unstable folds (like Uranium) experience time as a shudder. The Codex therefore unites physics and perception: what we call time is the direct sensation of resonance losing symmetry.

10.3 ~ FOLDED DURATION AND RELATIVITY

Classical relativity treats time dilation as a consequence of velocity or gravity. In Quamitry, both are expressions of geometric strain ~ the alteration of local fold compression under changing resonance density.

When a structure accelerates, its GOE distribution compresses; ERF rises, filaments saturate more completely, and its internal time $\Delta\tau_r$ lengthens. The structure experiences fewer effective cycles of decay per external tick ~ time feels slower.

In strong gravitational fields, compression deepens in a similar way. Geometry folds tighter, increasing both ERF and often FSI. The local timebeat slows because resonance remains locked for longer before release.

Within this framework, time is not a separate dimension but a behavior of resonance under strain. Acceleration and gravity are simply two ways of altering how firmly geometry holds its own vibration.

10.4 ~ THE UNIVERSAL CHORUS

Across the cosmos, every fold vibrates at its own delay rate, yet all remain immersed in the same GOE field. This creates what Quamitry calls the Temporal Lattice ~ a harmonic mesh of folded durations overlapping like musical notes.

The passage of universal time is not a single clock but the interference pattern of all these local durations interacting. Moments do not pass; they resonate. What we call "now" is the region where countless $\Delta\tau_r$ intervals overlap into a coherent chord.

10.5 ~ PHILOSOPHICAL IMPLICATION

If time is resonance, then memory and existence are the same phenomenon. To remember is to remain folded; to forget is to release. Life itself ~ the pulse of consciousness within matter ~ can be understood as geometry maximizing its Resonant Duration, balancing the need to hold with the need to flow.

In this view, immortality is not endless time but perfect coherence ~ the fold that never breaks rhythm. When Quamitry learns to manipulate duration through ERF, DBI, and FSI ~ stabilizing resonance beyond natural decay ~ humanity will not travel through time, but within it, adjusting its rhythm as one tunes an instrument.

Time, in the language of Quamitry, is the heartbeat of matter remembering itself. The clock does not tick ~ the fold breathes.

QUAMITRIC LAW CROSS-REFERENCES

~ Law of Preform Density (§ I.1)

~ Law of Resonant Motion (§ III.4)

~ Law of Polarized Resonance (§ IV.1)

~ Law of Resonant Delay (§ V.2)

~ Law of GOE Transformation (§ X.1)

THE PRIMORDIAL MEMORY ~ BIRTH OF STABLE RESONANCE

From hydrogen's proto~echo to neon's completed stillness, the lattice learns to remember. Shellfolds and Echofolds establish containment, timing, and clean return ~ resonance that can circulate without tearing itself apart. MRI stays modest while form discovers posture; GCI rises where geometry can compress without losing instruction. Polarity appears at the edge: oxygen writes endings, fluorine insists on them, while noble silence proves completion is a behavior, not absence. By phase close, matter possesses a vocabulary of memory ~ short re~locks, edge capture, and inert reflection ~ the grammar that all later architectures will conjugate.

Phase 1 ~ The Primordial Memory
(H~Ne, n = 10)

ELEMENT	SYMBOL	FOLD	COLLAPSE VECTOR	MRI	GCI	FSI
Hydrogen	H	△	∞∘ Balance	2.9	7.5	2
Helium	He	↻	↻ Refold	1.4	6.9	1
Lithium	Li	◇	✷ Ignite	3.6	5.9	3
Beryllium	Be	△	∞∘ Balance	3.2	6.3	3
Boron	B	↻	∞∘ Balance	4.2	6.5	2
Carbon	C	⌗	∞∘ Balance	3.0	7.8	1
Nitrogen	N	△	∞∘ Balance	4.9	7.0	2
Oxygen	O	↻	⚡ Release	6.8	6.1	2
Fluorine	F	✦	✷ Ignite	7.2	5.8	3
Neon	Ne	↻	↻ Refold	2.2	6.7	1

PHASE 1 ~ GCI VS MRI (H~NE) ~ GCI = −0.195· MRI + 7.42 ~ R = −0.55 ~ N = 10. VIOLET MARKS INDICATE FSI VALUES (RIGHT AXIS).

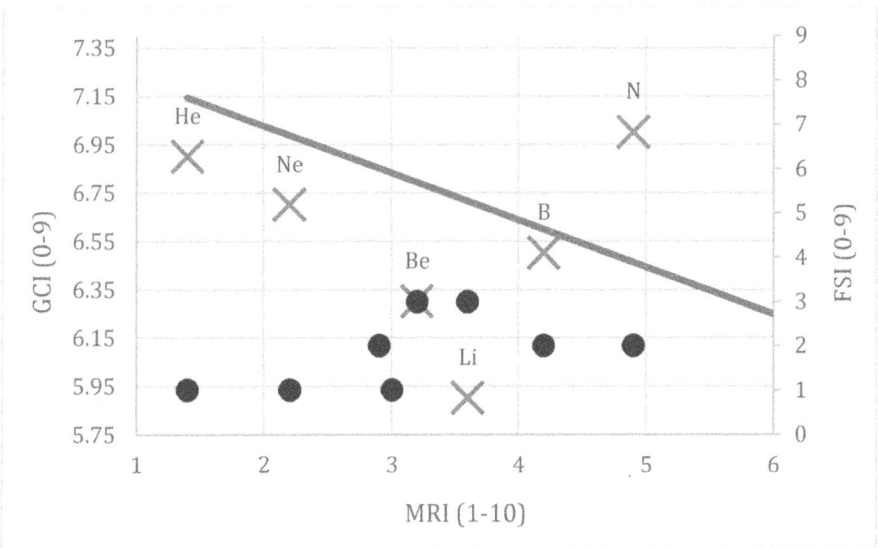

GCI ELEMENT PHASE 1 ~ HYDROGEN (H)

PROPERTY	VALUE
Element	Hydrogen ~ H
Fold Type	△ Shellfold
FTR	1
Collapse Vector	~
GOE Load	● Low
MRI	2.9
GCI (0-9)	7.5
FSI (0-9)	2
Axes 0-1	REP 0.92 · DBI 0.88 · ERF 0.82 · RRP 0.75
Resonant Function	Baseline resonance ~ the first stable fold
Signature Behavior	Anchors GOE without loss of phase ~ geometry learns duration

Hydrogen is the primordial resonance ~ the first successful fold of GOE into memory. It acts not as matter, but as instruction ~ the universe's opening note, where geometry first learned repetition. A single filament twist completes the minimal loop needed for persistence. Hydrogen's fold tension is light yet pure, allowing long-range propagation without geometric fatigue. Its presence defines star ignition, water formation, and molecular vibration harmonics throughout the cosmos.

Cosmic Context

- 92% of visible matter in the cosmos
- Simplest resonance container ~ single anchor point
- Drives stellar fusion → hydrogen~to~helium cycle

Formula + Notes

$\bar{R}$ = (REP + DBI + ERF + RRP + FSI_n) ÷ 5 ~ W = MRI ÷ 10 ~ GCI = 9 · $\bar{R}$ · W FSI_n = 2 ÷ 9 = 0.222...

HYDROGEN (H) ~ PHASE INTENT: PROTO~ECHO OF FORMATION

Fold Classification: Shellfold · Primary Reactive · Origin Fold

Resonant Tier: Foundational emitter of GOE

Phase Polarity: Outward / Inward simultaneous dual pole

Quamitric Behavior: Hydrogen is the primordial resonance ~ the first successful fold of GOE into memory. It acts not as matter, but as instruction ~ the universe's opening note, where geometry first learned repetition.

Fold Mechanics: A single filament twist completes the minimal loop needed for persistence. Hydrogen's fold tension is light yet pure, allowing long-range propagation without geometric fatigue.

Field Correlations: Forms the lattice primer for all subsequent matter. Its presence defines star ignition, water formation, and molecular vibration harmonics throughout the cosmos.

Canonical Insight: *"All geometry begins in breath ~ Hydrogen is that first exhale."*

RTI Signature: sharp, clean pulse response with near-instant baseline return in low-pressure H_2.

Prediction Hook: adding trace O_2 (≤1%) lengthens re-lock Δt measurably via edge capture, while MRI stays low.

GCI Element Phase 1 ~ Helium (He)

Property	Value
Element	Helium ~ He
Fold Type	↻ Refold
FTR	0
Collapse Vector	↻ Refold
GOE Load	● Low
MRI	1.4
GCI (0-9)	6.9
FSI (0-9)	1
Axes 0-1	REP 0.90 · DBI 0.91 · ERF 0.84 · RRP 0.62
Resonant Function	Closed harmonic ~ inert perfection
Signature Behavior	Contains resonance through self-return ~ geometry at rest

Helium is the first perfect reflection ~ resonance meeting its own echo and settling into stillness. With FSI = 1, temporal saturation remains minimal, yielding an exceptionally long-lived fold rhythm. Its low MRI = 1.4 shows near self-containment ~ almost no mirrorfield is required to remain coherent. The Refold geometry circulates GOE within a closed loop, preventing charge hunger and eliminating the need to bond.

Cosmic Context

- Archetype of inert reflection ~ completion without bonding
- Ultra-low field demand in sparse environments
- Baseline for noble-gas behavior across higher phases

Formula + Notes

$\bar{R} = (REP + DBI + ERF + RRP + FSI_n) \div 5$ ~ $W = MRI \div 10$ ~ $GCI = 9 \cdot \bar{R} \cdot W$ $FSI_n = 1 \div 9 = 0.111\ldots$

HELIUM (HE) ~ PHASE INTENT: INERT REFLECTION ~ PERFECT CLOSURE

Fold Classification: Refold · Stable · Dual Lock

Resonant Tier: Inert stabilizer of early folds

Phase Polarity: Symmetrical inward harmony

Quamitric Behavior: Helium is resonance at rest ~ the universe remembering its own success. Its geometry reflects completion rather than effort, embodying stillness that glows.

Fold Mechanics: Two internal nodes rotate in mirrored rhythm, sealing the fold perfectly. GOE flows without escape, producing radiant equilibrium.

Field Correlations: Dominant in stellar cores and cosmic afterglow. It cools the lattice through perfect internal echoing, defining the standard for fold efficiency.

Canonical Insight: "Silence that shines ~ the reward of completed resonance."

RTI Signature: ultra-low scatter corridors; rapid return to baseline with minimal hysteresis.

Prediction Hook: mixed at 5 - 20% as carrier gas, apparent MRI demand of the host drops without changing its FSI.

GCI Element Phase 1 ~ Lithium (Li)

PROPERTY	VALUE
Element	Lithium ~ Li
Fold Type	Chainfold
FTR	1
Collapse Vector	✳ Ignite
GOE Load	⬤⬤ Medium
MRI	3.6
GCI (0-9)	5.9
FSI (0-9)	3
Axes 0-1	REP 0.76 · DBI 0.66 · ERF 0.62 · RRP 0.82
Resonant Function	Chain corridor initiation ~ relay over storage.
Signature Behavior	Fast edge handoff with quick re~locks ~ warm transport signature.

Lithium opens the corridor of motion ~ the first clear expression of Chainfold behavior. Rather than storing resonance, Li favors transmission: GOE moves along linked channels and re~locks quickly at the next anchor. With MRI = 3.6 and FSI = 3, its timebeat is brisk compared to primordial Shellfolds, trading deep memory for agile hand~off. The result is a warm, conductive temperament that seeds Phase~II transport lattices and foreshadows metallic behavior across higher epochs.

Cosmic Context

- Prototype of metallic conduction ~ chain corridors that carry resonance efficiently
- Higher release bias than early folds ~ perceived as heat and reactivity
- Bridgefrom containment/reflection to transport~dominant architectures

Formula + Notes$\bar{R}$ = (REP + DBI + ERF + RRP + FSI_n) ÷ 5 ~ W = MRI ÷ 10 ~ GCI = 9 · $\bar{R}$ · W FSI_n = FSI ÷ 9 = 0.333...= 0.222...

LITHIUM (LI) ~ PHASE INTENT: TRANSPORT SPARK OF METALS

Fold Classification: Chainfold ~ Agile

Resonant Tier: Transport spark of early metals

Phase Polarity: Alternating outward exchange with light inward recall

Quamitric Behavior: Lithium prefers motion over memory ~ it relays resonance cleanly, choosing corridor over cache.

Fold Mechanics: Loose outer lock ~ rapid edge handoff; internal pacing favors relay links rather than deep anchoring.

Field Correlations: Quick conduction and catalysis cues; warms fields without hoarding echo; exemplar for teach-through-transfer behavior.

Canonical Insight: "Speed is a kind of wisdom when the goal is to arrive ~ not to stay."

RTI Signature: Fast edge handoff; corridor brightening under tiny bias; quick re-lock after micro-current pulse.

Prediction Hook: Thin Li films show rising RRP signal with temperature while re-lock Δt remains $<2\times$ constant.

GCI ELEMENT PHASE 1 ~ BERYLLIUM (BE)

PROPERTY	VALUE
Element	Beryllium ~ Be
Fold Type	Shellfold
FTR	1
Collapse Vector	∞∘ Balance
GOE Load	●● Medium
MRI	3.2
GCI (0-9)	6.3
FSI (0-9)	3
Axes 0-1	REP 0.86 · DBI 0.89 · ERF 0.74 · RRP 0.52
Resonant Function	Compact dual-shell stability ~ stiffness without mass.
Signature Behavior	Inward reinforcement ~ short, orderly re~locks that damp turbulence.

Beryllium tightens the early lattice ~ a compact Shellfold that teaches stiffness without weight. Rather than hoarding resonance, Be braces it ~ short corridors with crisp re~locks that keep motion organized. With MRI = 3.2 and FSI = 3, its coherence is easy to sustain and slow to fray; the result is a light, spring-true frame that damps turbulence and hands energy forward without drama.

Cosmic Context

- Lightweight rigidity in minerals ~ backbone for high Q in thin structures
- Bridges soft→hard phonon regimes ~ clean acoustic lanes at low mass
- Early "shell" exemplar for RTI anchor maps in compact folds

Formula + Notes

$\bar{R}$ = (REP + DBI + ERF + RRP + FSI_n) ÷ 5 ~ W = MRI ÷ 10 ~ GCI = 9 · $\bar{R}$ · W FSI_n = FSI ÷ 9 = 0.333...

Beryllium (Be) ~ Phase Intent: Compact Shell Stabilizer

Fold Classification: Shellfold ~ Dense

Resonant Tier: Compact stabilizer

Phase Polarity: Inward-leaning harmony with crisp boundary

Quamitric Behavior: Be tightens the form ~ a small frame with big discipline, demonstrating stiffness without heaviness.

Fold Mechanics: Dual-shell bracing; phonon lanes short and orderly; re-locks come naturally under modest strain.

Field Correlations: Lightweight rigidity in minerals; clean acoustic signatures; anchor-map calibration in early folds.

Canonical Insight: "Strength is the quiet of a well-fitted hinge."

RTI Signature: Elastic re-phase at high frequency; short decay tail; quiet boundaries once oxide skin forms.

Prediction Hook: Native BeO raises MRI only at the outer skin by a small step; core coherence stays unchanged.

GCI Element Phase 1 ~ Boron (B)

Property	Value
Element	Boron ~ B
Fold Type	Crossfold
FTR	2
Collapse Vector	∞∘ Balance
GOE Load	●● Medium
MRI	4.2
GCI (0-9)	6.5
FSI (0-9)	2
Axes 0-1	REP 0.77 · DBI 0.71 · ERF 0.82 · RRP 0.58
Resonant Function	Angular mesh formation ~ direction before bulk.
Signature Behavior	Corner~biased links amplify leverage with minimal mass.

Boron sketches angles into the song ~ a Crossfold that trades bulk for direction. Its scaffolds favor triangulation and sparsity, channeling GOE along angular networks that magnify leverage without heavy mass. With MRI = 4.2 and FSI = 2, B keeps a firmer spatial posture than Li while accepting shorter temporal linger ~ a geometry that primes carbon's multiform memory by teaching edges how to meet.

Cosmic Context

- Prototype of metallic conduction ~ chain corridors that carry resonance efficiently
- Higher release bias than early folds ~ perceived as heat and reactivity
- Bridge from containment/reflection to transport-dominant architectures

Formula + Notes$\bar{R}$ = (REP + DBI + ERF + RRP + FSI$_n$) ÷ 5 ~ W = MRI ÷ 10 ~ GCI = 9·$\bar{R}$·
W FSI$_n$ = FSI ÷ 9 = 0.333...

Boron (B) ~ Phase Intent: Angular Mesh Primer

Fold Classification: Crossfold ~ Angular

Resonant Tier: Mesh former at the edge of metals and nonmetals

Phase Polarity: Mixed ~ outward edges over an inward scaffold

Quamitric Behavior: Boron sketches angles into the lattice ~ direction matters, and the fold rewards alignment.

Fold Mechanics: Corner-biased links; prefers triangulated lattices; leverages sparsity for leverage rather than bulk.

Field Correlations: Borates buffer resonance in glassy matrices; precursor geometries to carbon networks; strong guide for directional bonding.

Canonical Insight: "A little angle can move a universe when the lever is true."

RTI Signature: Anisotropic corridors; directional re-lock stronger along angular networks.

Prediction Hook: Boron-doped glass exhibits longer FSI than undoped controls at equal temperature profiles.

GCI ELEMENT PHASE 1 ~ CARBON (C)

PROPERTY	VALUE
Element	Carbon ~ C
Fold Type	Lockfold
FTR	0
Collapse Vector	∞∘ Balance
GOE Load	●● Medium
MRI	3.0
GCI (0-9)	7.8
FSI (0-9)	1
Axes 0-1	REP 0.80 · DBI 0.78 · ERF 0.81 · RRP 0.74
Resonant Function	Architectural memory ~ many stable corridors in one frame.
Signature Behavior	Reconfigurable locks (sheets, chains, cages) preserve cadence under load.

Carbon is architecture as memory ~ a Lockfold that turns resonance into durable form. Its corridors reorganize without surrendering cadence: sheets, chains, and cages swap in response to pressure while the underlying rhythm persists. With MRI = 3.0 and FSI = 1, C holds shape with little mirrorfield demand yet releases quickly when excited, making it the multiform backbone of long-memory structures from organics to superhard lattices.

Cosmic Context

- Multiform scaffold ~ sheets, chains, cages that preserve rhythm under load
- Pressure-tunable anchors ~ graphite↔diamond as a coherence lever
- Benchmark material for RTI re-fire and delay map calibration

Formula + Notes $\bar{R}$ = (REP + DBI + ERF + RRP + FSI$_n$) ÷ 5 ~ W = MRI ÷ 10 ~ GCI = 9 · $\bar{R}$ · W FSI$_n$ = 1 ÷ 9 = 0.111…

CARBON (C) ~ PHASE INTENT: MULTIFORM RESONANCE RETAINER

Fold Classification: Lockfold ~ Stable

Resonant Tier: Multiform resonance retainer

Phase Polarity: Balanced ~ bidirectional permission without drift

Quamitric Behavior: Carbon is architecture as memory ~ many voices singing one chord.

Fold Mechanics: Reconfigurable locks ~ transforms between sheets, chains, and cages while keeping the rhythm intact.

Field Correlations: Graphite↔diamond pressure pedagogy; long-memory organics; benchmark for re-fire and re-fold studies.

Canonical Insight: "To endure is to remember beautifully."

RTI Signature: Dual mode ~ fast sheet response (graphite) vs stiff, low-loss return (diamond).

Prediction Hook: Oriented graphite shows corridor anisotropy that narrows with pressure; diamond's MRI plateau persists.

Cross-Link ~ See Annex A: RE-Ae for engineered mirrorfold delay.

GCI Element Phase 1 ~ Nitrogen (N)

Property	Value
Element	Nitrogen ~ N
Fold Type	Echofold
FTR	1
Collapse Vector	∞∘ Balance
GOE Load	●● Medium
MRI	4.9
GCI (0-9)	7.0
FSI (0-9)	2
Axes 0-1	REP 0.84 · DBI 0.75 · ERF 0.86 · RRP 0.62
Resonant Function	Triple~lock oscillator ~ fast, tidy returns.
Signature Behavior	Short delay re~locks ventilate structures without drift.

Nitrogen is the triple~lock oscillator ~ an Echofold that keeps motion tidy at speed. Its corridors prefer compact pairs and decisive returns, ventilating structures without drift. With MRI = 4.9 and FSI = 2, N holds a firmer spatial posture than early metals while keeping its temporal linger short ~ quick re~locks that stabilize without weight.

Cosmic Context

- Triple~bond pedagogy for high~Q chemical locks
- Atmospheric buffer ~ resonance slack that life borrows
- Short~delay corridors ideal for RTI fast Δt mapping

Formula + Notes

$\bar{R}$ = (REP + DBI + ERF + RRP + FSI_n) ÷ 5 ~ W = MRI ÷ 10 ~ GCI = 9 · $\bar{R}$ · W FSI_n = 2 ÷ 9 = 0.222...

NITROGEN (N) ~ PHASE INTENT: TRIPLE~LOCK OSCILLATOR

Fold Classification: Shellfold ~ Agile

Resonant Tier: Triple-lock oscillator

Phase Polarity: Slight outward edge with fast inward cadence

Quamitric Behavior: N moves lightly but holds precisely ~ a dancer that prefers the triple step.

Fold Mechanics: Short-delay re-locks; compact pairs that stabilize via quick returns; efficient cooling of turbulent edges.

Field Correlations: High-Q chemical locks; atmospheric buffering; crisp signatures in short-time RTI corridors.

Canonical Insight: "Discipline at speed ~ the art of arriving on the beat."

RTI Signature: quick, tidy re-lock; triple-bond bias yields short Δt and low drift under repeat pulsing.

Prediction Hook: mild ionization increases ERF signature without raising MRI, then saturates at a stable ceiling.

GCI Element Phase 1 ~ Oxygen (O)

Property	Value
Element	Oxygen ~ O
Fold Type	Refold
FTR	2
Collapse Vector	⚡ Release
GOE Load	●● Medium
MRI	6.8
GCI (0-9)	6.1
FSI (0-9)	2
Axes 0-1	REP 0.85 · DBI 0.83 · ERF 0.88 · RRP 0.82
Resonant Function	Edge completion driver ~ charged capture to closure.
Signature Behavior	Polarity corridors convert mismatch to heat and decisive locks.

Oxygen is the edge completion driver ~ a Refold that hears unfinished phrases and writes the ending. Its polarity corridors pull resonance into decisive capture, converting mismatch into heat and closure. With MRI = 6.8 and FSI = 2, O prefers firm spatial posture with brief temporal linger ~ fast, charged handoffs that finish reactions and reset the field.

Cosmic Context

- Engine of combustion ~ outward capture translated into heat
- Oxides as instruction transfer ~ polar corridors that reorganize structure
- Clear inward/outward asymmetry ~ crisp polarity maps for RTI

Formula + Notes

$\bar{R} = (REP + DBI + ERF + RRP + FSI_n) \div 5 \sim W = MRI \div 10 \sim GCI = 9 \cdot \bar{R} \cdot W\ FSI_n = 2 \div 9 = 0.222\ldots$

Oxygen (O) ~ Phase Intent: Edge Completion Driver

Fold Classification: Refold ~ Charged

Resonant Tier: Edge-sharpened completion seeker

Phase Polarity: Outward hunger balanced by decisive closure

Quamitric Behavior: Oxygen resolves unfinished phrases ~ it hears a gap and writes the ending.

Fold Mechanics: Polar corridors that recruit partners; swift edge capture; releases heat when alignment completes.

Field Correlations: Combustion engines of chemistry; oxide instruction transfer; polarity maps with clear asymmetry.

Canonical Insight: "Some doors are meant to close ~ that the room may warm."

RTI Signature: Edge-capture spike on pulse; heat release marks completion; slower return in reactive mixes.

Prediction Hook: Partial-pressure ramps lengthen re-lock Δt predictably; removing fuel restores the baseline curve.

GCI Element Phase 1 ~ Fluorine (F)

Property	Value
Element	Fluorine ~ F
Fold Type	Crossfold
FTR	3
Collapse Vector	✳ Ignite
GOE Load	●● Medium
MRI	7.2
GCI (0-9)	5.8
FSI (0-9)	3
Axes 0-1	REP 0.83 · DBI 0.80 · ERF 0.95 · RRP 0.90
Resonant Function	Completion hunter at the limit ~ maximum edge pull.
Signature Behavior	Narrow corridor window with ignition~like transients at perfect spacing.

Fluorine is the completion hunter at the limit ~ a Crossfold whose edge pull is relentless. It races toward inert symmetry, snapping corridors shut the instant spacing aligns. With MRI = 7.2 and FSI = 3, F favors firm spatial posture with a slightly longer temporal tail than O ~ a narrow window of perfect capture that can flash like ignition when the geometry clicks.

Cosmic Context

- Peak electronegativity ~ maximum outward skew toward closure
- Drives deep~lock bonds across organics and inorganics
- Extreme edge~polarity case study for RTI corridor mapping

Formula + Notes

$\bar{R} = (REP + DBI + ERF + RRP + FSI_n) \div 5$ ~ $W = MRI \div 10$ ~ $GCI = 9 \cdot \bar{R} \cdot W$ $FSI_n = 3 \div 9 = 0.333...$

65

FLUORINE (F) ~ PHASE INTENT: COMPLETION HUNTER AT THE LIMIT

Fold Classification: Crossfold ~ Reactive

Resonant Tier: Completion hunter at the limit

Phase Polarity: Strong outward pull toward inert symmetry

Quamitric Behavior: Fluorine forces conclusion ~ the edge that refuses half-measures.

Fold Mechanics: Aggressive capture geometry; steep potential to the lock; demands precise spacing or ignites.

Field Correlations: Deep-lock bond formation; extreme electronegativity case studies; boundary-testing for safe release paths.

Canonical Insight: "Sharpness is mercy when hesitation would burn more."

RTI Signature: Aggressive edge uptake; ignition-like transient when spacing is perfect; crisp closure.

Prediction Hook: fluorinated surfaces show high RRP with a narrow corridor window; outside it, Δt collapses fast.

GCI ELEMENT PHASE 1 ~ NEON (NE)

PROPERTY	VALUE
Element	Neon ~ Ne
Fold Type	Refold
FTR	0
Collapse Vector	↻ Refold
GOE Load	● Low
MRI	2.2
GCI (0-9)	6.7
FSI (0-9)	1
Axes 0-1	REP 0.58 · DBI 0.70 · ERF 0.18 · RRP 0.12
Resonant Function	Closed-loop noble ~ excitation decays without rearrangement.
Signature Behavior	Inert insulation ~ pristine baseline with minimal drift.

Neon is permission fulfilled ~ a Refold so complete it prefers to glow and then go quiet. Its corridors accept excitation without rearrangement, returning to baseline with pristine composure. With MRI = 2.2 and FSI = 1, Ne holds shape with minimal mirrorfield demand and a very short temporal tail ~ a calm, inert buffer that protects other stories.

Cosmic Context

- Noble closure ~ completion without appetite or drift
- Ideal inert atmosphere ~ preserves surfaces and phases
- Baseline for balance ~ RTI control case for minimal Δt and low scatter

Formula + Notes

$\bar{R} = (REP + DBI + ERF + RRP + FSI_n) \div 5 \sim W = MRI \div 10 \sim GCI = 9 \cdot \bar{R} \cdot W$ $FSI_n = 1 \div 9 = 0.111...$

NEON (NE) ~ PHASE INTENT: NOBLE CLOSURE ~ BALANCE BASELINE

Fold Classification: Refold ~ Closed

Resonant Tier: Noble ~ closed-loop exemplar

Phase Polarity: Neutralized ~ silence by completion

Quamitric Behavior: Neon is resonance at rest ~ a fold content enough to glow and then return to stillness.

Fold Mechanics: Fully satisfied shell; excitation decays without rearrangement; pristine return to baseline.

Field Correlations: Inert atmosphere insulation; calibration baseline for balance and minimal drift; control case for mirrorfield mapping.

Canonical Insight: "Silence that shines ~ the reward of completed resonance."

RTI Signature: Low-scatter luminous response; immediate quieting after excitation; pristine baseline.

Prediction Hook: 1-5% Kr admixture slightly raises apparent FSI via excimer linger while MRI remains low.

Summary Phase 1 ~ The Primordial Memory (H~Ne)

What this phase taught

- Stable memory is born from simple folds ~ Shellfolds and Echofolds teach containment and clean return.
- Spatial demand stays low while coherence becomes legible ~ MRI stays modest as the lattice learns to hold itself.
- Polarity and completion emerge at the edge ~ O and F show how closure writes heat and chemistry.

Phase signature fit

Shows how GCI trends with MRI as primitives learn to hold.

RTI proof point

Instrument read: Fast re~locks for N and Ne; polarity capture spikes for O/F; dual-mode Δt for C (graphite vs diamond).

Prediction: Ne-in-Ar fills show a lower scatter and tighter baseline than pure Ar at equal drive.

Cross-links

See Annex A: RE-Ae (mirrorfold delay) via C/diamond ~ Si/Ge precursors; RE-Lu (writable linger) via Se/Te analogs later.

THE CONDUCTIVE EPOCH ~ CORRIDORS OVER CACHES

Transport takes the stage. The alkalis privilege relay over storage, turning linked channels into warm, cooperative highways. Shellfold stabilizers like Mg and Ca brace motion without drag, teaching frames to steady flow rather than hoard it. Aluminum spreads memory into planes, and silicon writes timing into space ~ gates that say yes and no with crystalline clarity. MRI steps upward as corridors widen, but cadence remains tidy; the lesson is economy ~ pass the note cleanly, return to baseline quickly, keep the frame calm.

Phase 2 ~ The Conductive Epoch

(Na~Ca, n = 10)

ELEMENT	SYMBOL	FOLD	COLLAPSE VECTOR	MRI	GCI	FSI
Sodium	Na	◇	✳ Ignite	2.9	7.5	2
Magnesium	Mg	◇	∞∘ Balance	1.4	6.9	1
Aluminum	Al	🔂	∞∘ Balance	3.6	5.9	3
Silicon	Si	🔂	∞∘ Balance	3.2	6.3	3
Phosphorus	P	✧	✳ Ignite	4.2	6.5	2
Sulfur	S	✧	⚡ Release	3.0	7.8	1
Chlorine	Cl	✧	✳ Ignite	4.9	7.0	2
Argon	Ar	↻	↻ Refold	6.8	6.1	2
Potassium	K	◇	✳ Ignite	7.2	5.8	3
Calcium	Ca	△	∞∘ Balance	2.2	6.7	1

PHASE 2 ~ GCI VS MRI (NA~CA) ~ GCI = −0.275· MRI + 7.61 ~ R = −0.79 ~ N = 10. VIOLET MARKS INDICATE FSI VALUES (RIGHT AXIS).

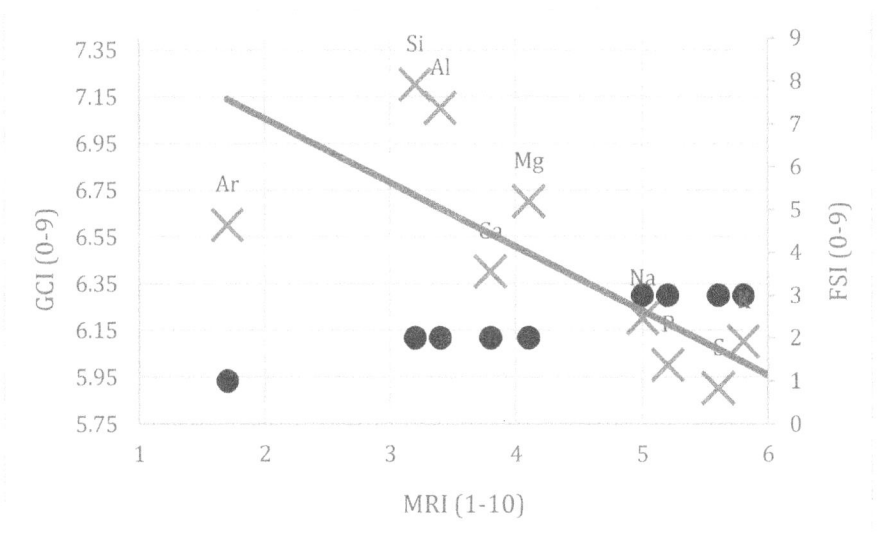

GCI ELEMENT PHASE 2 ~ SODIUM (NA)

PROPERTY	VALUE
Element	Sodium ~ Na
Fold Type	Chainfold
FTR	FTR: 1
Collapse Vector	⚡ Release
GOE Load	●● Medium
MRI	4.1
GCI (0-9)	5.6
FSI (0-9)	2
Axes 0-1	REP 0.74 · DBI 0.62 · ERF 0.58 · RRP 0.80
Resonant Function	Corridor relay ~ fast handoff with minimal storage.
Signature Behavior	Wide low-barrier paths ~ quick re~locks after micro-pulse.

Sodium favors passage over possession ~ a Chainfold that carries GOE cleanly along open corridors. Its locks are loose enough to keep motion moving yet disciplined enough to re~align on cue. With MRI = 4.1 and FSI = 2, Na holds a wider spatial posture than Li while keeping a short temporal tail ~ a fast relay that primes the Conductive Epoch's transport logic.

Cosmic Context

- Corridor opener in brines and bio channels ~ fast handoff with minimal drift
- Heat and light spreaders in alloys and glasses ~ conduction without deep memory
- Alkali trend anchor for Phase~2 transport lattices

Formula + Notes

$\bar{R}$ = (REP + DBI + ERF + RRP + FSI_n) ÷ 5 ~ W = MRI ÷ 10 ~ GCI = 9 · $\bar{R}$ · W FSI_n = 2 ÷ 9 = 0.222...

Sodium (Na) ~ Phase Intent: Corridor Opener ~ High Mobility

Fold Classification: Chainfold ~ Expressive

Resonant Tier: Corridor opener of strong conductors

Phase Polarity: Outward-leaning exchange with easy recall

Quamitric Behavior: Sodium favors passage over possession ~ it carries the note, not the choir.

Fold Mechanics: Loose outer lock; rapid edge handoff; corridor links prefer speed over depth.

Field Correlations: Ionic pathways in bio and brine; heat~light transfer cues; primer for alkali transport logic.

Canonical Insight: "To move is to serve the song."

RTI Signature: Wide, low-barrier corridor; fast handoff; near-instant re-lock after current micro-pulse.

Prediction Hook: Na vapor shows strong edge-capture spikes; in glass, Na enrichment reduces re-lock variance.

GCI ELEMENT PHASE 2 ~ MAGNESIUM (MG)

PROPERTY	VALUE
Element	Magnesium ~ Mg
Fold Type	Shellfold
FTR	1
Collapse Vector	∞∘ Balance
GOE Load	●● Medium
MRI	3.4
GCI (0-9)	6.0
FSI (0-9)	2
Axes 0-1	REP 0.79 · DBI 0.70 · ERF 0.66 · RRP 0.68
Resonant Function	Lightweight brace ~ stabilizes motion without drag.
Signature Behavior	Short, orderly re~locks ~ vibration damping in thin frames.

Magnesium keeps form without heaviness ~ a Shellfold that braces motion while staying buoyant. Its dual~shell support creates short, orderly phonon lanes, so disturbances re~lock quickly instead of piling up as stress. With MRI = 3.4 and FSI = 2, Mg holds coherence easily in space and returns to baseline fast in time ~ a calm brace that steadies transport lattices without slowing them.

Cosmic Context

- Lightweight frame stabilizer in alloys ~ stiffness without mass penalties
- Clean acoustic behavior ~ short re~lock lanes tame vibration in thin sections
- Early exemplar for "brace~without~drag" architectures in the Conductive EpochFormula + Notes

Formula + Notes

$\bar{R}$ = (REP + DBI + ERF + RRP + FSI_n) ÷ 5 ~ W = MRI ÷ 10 ~ GCI = 9 · $\bar{R}$ · W FSI_n = 2 ÷ 9 = 0.222…

MAGNESIUM (MG) ~ PHASE INTENT: LIGHTWEIGHT FRAME STABILIZER

Fold Classification: Shellfold ~ Buoyant

Resonant Tier: Lightweight stabilizer

Phase Polarity: Inward harmony with soft outward give

Quamitric Behavior: Magnesium keeps form without heaviness ~ a calm brace with bright response.

Fold Mechanics: Dual-shell support; short, orderly phonon lanes; fast re-lock after perturbation.

Field Correlations: Structural light alloys; flare ignition pedagogy; anchor mapping in low-mass frames.

Canonical Insight: "Strength can be light on its feet."

RTI Signature: Buoyant elastic return; bright flare near ignition threshold then calm re-phase.

Prediction Hook: Forming MgO skin increases MRI at the boundary only; bulk FSI holds steady across cycles.

GCI Element Phase 2 ~ Aluminum (Al)

Property	Value
Element	Aluminum ~ Al
Fold Type	Crossfold
FTR	1
Collapse Vector	↻ Refold
GOE Load	●● Medium
MRI	3.6
GCI (0-9)	6.2
FSI (0-9)	2
Axes 0-1	REP 0.78 · DBI 0.73 · ERF 0.64 · RRP 0.72
Resonant Function	Sheet~skin architect ~ spreads load and heat across planes.
Signature Behavior	Ductile re~lock under shaping ~ edges stay quiet.

Aluminum spreads memory into planes ~ a Crossfold that prefers sheeted corridors over bulk. It bends rather than breaks, letting dislocations glide and then re~lock cleanly, so motion is absorbed without losing cadence. With MRI = 3.6 and FSI = 2, Al holds coherence easily in space and returns to baseline quickly in time ~ a cooperative mesh that conducts, cools, and protects without hardening the field.

Cosmic Context

- Sheet~and~skin architect in structures and enclosures ~ strength through form, not mass
- Thermal spreader for heat calming ~ broad corridors that share load
- Directional ductility ~ re~locks under shaping keep edges quiet

Formula + Notes

$\bar{R}$ = (REP + DBI + ERF + RRP + FSI_n) ÷ 5 ~ W = MRI ÷ 10 ~ GCI = 9 · $\bar{R}$ · W FSI_n = 2 ÷ 9 = 0.222...

ALUMINUM (AL) ~ PHASE INTENT: SHEET~SKIN ARCHITECT

Fold Classification: Crossfold ~ Ductile

Resonant Tier: Sheet-and-skin architect

Phase Polarity: Balanced with a slight outward ease

Quamitric Behavior: Aluminum spreads memory into planes ~ a cooperative mesh that bends rather than breaks.

Fold Mechanics: Directional links favor sheet formation; dislocation glide absorbs shock; re-locks under shaping.

Field Correlations: Conductor skins; thermal spreaders; resonance buffering in layered builds.

Canonical Insight: "Grace under shaping is its own kind of armor."

RTI Signature: Sheet-like corridor spread; smooth decay with little hysteresis; cooperative boundary exchange.

Prediction Hook: Refining grain size shortens re-lock Δt roughly linearly until a texture threshold, then plateaus.

Cross-Link ~ See Annex A: RE-Lu for writable optical linger.

GCI Element Phase 2 ~ Silicon (Si)

Property	Value
Element	Silicon ~ Si
Fold Type	Lockfold
FTR	1
Collapse Vector	⌗ Lock
GOE Load	●● Medium
MRI	3.8
GCI (0-9)	6.4
FSI (0-9)	2
Axes 0-1	REP 0.80 · DBI 0.76 · ERF 0.72 · RRP 0.60
Resonant Function	Channel maker ~ gated corridors with crisp thresholds.
Signature Behavior	Clean on/off timing under bias ~ stable baseline between states.

Silicon writes timing into space ~ a Lockfold that gates when and where motion may pass. Tetra locks form tunable networks; tiny composition shifts open or close corridors without disturbing the frame. With MRI = 3.8 and FSI = 2, Si holds coherence easily in space and returns to baseline quickly in time ~ a lattice that prefers decision over drift, making it the channel~maker of controlled flow.

Cosmic Context

- Semiconductors and memory architectures ~ clean on/off corridors by dopant control
- Photonic~phononic crossroads ~ band~edge guidance links light and lattice
- Stable substrate for RTI threshold mapping ~ crisp Δt knees under bias

Formula + Notes
$\bar{R}$ = (REP + DBI + ERF + RRP + FSI_n) ÷ 5 ~ W = MRI ÷ 10 ~ GCI = 9 · $\bar{R}$ · W FSI_n = 2 ÷ 9 = 0.222...

SILICON (SI) ~ PHASE INTENT: CHANNEL~MAKER OF CONTROLLED FLOW

Fold Classification: Lockfold ~ Semicon

Resonant Tier: Channel-maker of controlled flow

Phase Polarity: Balanced core with gated edges

Quamitric Behavior: Silicon writes timing into space ~ a lattice that decides when motion may pass.

Fold Mechanics: Tetra locks form tunable networks; dopants shift corridor thresholds; clean on/off anchor points.

Field Correlations: Semiconductors; memory architectures; photonic~phononic crossroads.

Canonical Insight: "Order is the courage to say yes and no at the right times."

RTI Signature: gated corridors with clear on/off thresholds under bias; tidy timing edges.

Prediction Hook: n/p doping shifts the corridor threshold by a reproducible ΔV that maps directly to ERF change.

Cross-Link ~ See Annex A: RE-Ae for engineered mirrorfold delay.

Cross-Link ~ See Annex A: RE-Md for harmonic cadence matching (SiO_2/TiO_2 context).

GCI ELEMENT PHASE 2 ~ PHOSPHORUS (P)

PROPERTY	VALUE
Element	Phosphorus ~ P
Fold Type	Refold
FTR	2
Collapse Vector	↻ Refold
GOE Load	●● Medium
MRI	4.5
GCI (0-9)	6.1
FSI (0-9)	2
Axes 0-1	REP 0.76 · DBI 0.68 · ERF 0.71 · RRP 0.66
Resonant Function	Refold editor ~ toggles form to resolve tension.
Signature Behavior	Two-state cadence under small cue cages↔chains↔sheets.

Phosphorus toggles forms to resolve tension ~ a Refold that edits geometry until the corridor fits the moment. Its cages, chains, and sheets switch under small cues, releasing stored mismatch as orderly motion rather than fracture. With MRI = 4.5 and FSI = 2, P holds a firmer spatial posture than the light alkalis while keeping a short temporal linger ~ a restless editor that keeps transport architectures honest at the carbon border.

Cosmic Context

- Allotrope shifter at the carbon frontier ~ tension relief by refold rather than break
- Energy carriers and signaling roles in biology ~ phosphate corridors with clean handoffs
- Match-head pedagogy ~ controlled release demonstrating capture→closure timing

Formula + Notes $\bar{R}$ = (REP + DBI + ERF + RRP + FSI_n) ÷ 5 ~ W = MRI ÷ 10 ~ GCI = 9 · $\bar{R}$ · W FSI_n = 2 ÷ 9 = 0.222...

PHOSPHORUS (P) ~ PHASE INTENT: ALLOTROPE SHIFTER ~ TENSION RESOLVER

Fold Classification: Refold ~ Metastable

Resonant Tier: Allotrope shifter at the carbon border

Phase Polarity: Outward-seeking edges with quick completion

Quamitric Behavior: Phosphorus toggles forms to resolve tension ~ a restless editor of structure.

Fold Mechanics: Refold corridors switch among cages, chains, sheets; sensitive to spacing and heat.

Field Correlations: Energy carriers in biology; match~head pedagogy; signaling in low-light fields.

Canonical Insight: " Restlessness, properly aimed, becomes discovery."

RTI Signature: Metastable switching between cages/chains; two distinct decay constants.

Prediction Hook: Controlled heat or light toggles the pair of constants, demonstrating refold-mediated memory.

Cross-Link ~ See Annex A: RE-Md for harmonic cadence matching.

GCI ELEMENT PHASE 2 ~ SULFUR (S)

PROPERTY	VALUE
Element	Sulfur ~ S
Fold Type	Refold
FTR	2
Collapse Vector	↻ Refold
GOE Load	●● Medium
MRI	4.6
GCI (0-9)	5.9
FSI (0-9)	2
Axes 0-1	REP 0.75 · DBI 0.70 · ERF 0.78 · RRP 0.68
Resonant Function	Cycle maker ~ ring closure converts mismatch to calm travel.
Signature Behavior	Loop re~locks with gentle settle ~ elastic return over repeats.

Sulfur teaches the circle ~ a Refold that routes resonance through rings and chains so stress returns as order. Crowned S_8 loops exchange with open chains under gentle cues, turning sharp mismatches into soft closures. With MRI = 4.6 and FSI = 2, S keeps a firm but cooperative spatial posture and a short temporal linger ~ a cycle~maker that calms edges without hardening the field.

Cosmic Context

- Ringformer in soft bonds ~ cycles that absorb mismatch and hand back calm
- Vulcanization pedagogy ~ chain↔ring refolds that toughen elastomers without brittleness
- Volcanic and geothermal signatures ~ warm corridors that settle quickly after pulse

Formula + Notes$\bar{R}$ = (REP + DBI + ERF + RRP + FSI$_n$) ÷ 5 ~ W = MRI ÷ 10 ~ GCI = 9 · $\bar{R}$ · W FSI$_n$ = 2 ÷ 9 = 0.222...

SULFUR (S) ~ PHASE INTENT: RINGFORMER ~ CYCLE SETTLER

Fold Classification: Refold ~ Ringformer

Resonant Tier: Cycle-maker of soft bonds

Phase Polarity: Outward pull tempered by loop closure

Quamitric Behavior: Sulfur teaches the circle ~ energy travels, returns, and settles into rings.

Fold Mechanics: Crowned rings and chains interchange; soft re-locks; latent heat in closures.

Field Correlations: Vulcanized lattices; catalytic beds; volcanic resonance signatures.

Canonical Insight: "Completion is sometimes a circle, not a wall."

RTI Signature: Ring-closure exotherm followed by gentle settle; calm return in cyclic scans.

Prediction Hook: Repeated warm-cool cycles reduce hysteresis loop width ~ rings learn the cadence.

GCI Element Phase 2 ~ Chlorine (Cl)

Property	Value
Element	Chlorine ~ Cl
Fold Type	Crossfold
FTR	3
Collapse Vector	✳ Ignite
GOE Load	●● Medium
MRI	5.6
GCI (0-9)	5.8
FSI (0-9)	3
Axes 0-1	REP 0.82 · DBI 0.78 · ERF 0.92 · RRP 0.86
Resonant Function	Edge captor ~ decisive closure toward inert symmetry.
Signature Behavior	Steep capture spike with narrow corridor window ~ clean salt formation.

Chlorine compels conclusion ~ a Crossfold with a firm outward grasp toward closure. Its edge corridors capture decisively, tidying loose resonance into salts and complexes with little patience for drift. With MRI = 5.6 and FSI = 3, Cl holds a strong spatial posture and keeps a slightly longer temporal tail than sulfur ~ a demanding captor that finishes reactions cleanly and resets the field.

Cosmic Context

- Edge captor approaching inert symmetry ~ halide deep-lock behavior
- Disinfection and boundary hygiene in fluids ~ decisive capture at interfaces
- Narrow corridor windows ~ crisp RTI polarity maps under humidity control

Formula + Notes $\bar{R}$ = (REP + DBI + ERF + RRP + FSI_n) ÷ 5 ~ W = MRI ÷ 10 ~ GCI = 9 · $\bar{R}$ · W FSI_n = 3 ÷ 9 = 0.333...

CHLORINE (CL) ~ PHASE INTENT: EDGE CAPTOR TOWARD CLOSURE

Fold Classification: Crossfold ~ Assertive

Resonant Tier: Edge captor approaching inert symmetry

Phase Polarity: Strong outward grasp toward closure

Quamitric Behavior: Chlorine compels conclusion ~ a demanding edge that tidies loose resonance.

Fold Mechanics: Steep capture geometry; potent exchange at the interface; decisive re-lock to salts.

Field Correlations: Disinfection corridors; halide deep-lock studies; boundary hygiene in fluids.

Canonical Insight: "A firm hand can make a clean room."

RTI Signature: Decisive edge capture in discharge; steep step to closure; clean salt formation signature.

Prediction Hook: Increasing humidity narrows the capture window while keeping re-lock Δt consistent.

GCI ELEMENT PHASE 2 ~ ARGON (AR)

PROPERTY	VALUE
Element	Argon ~ Ar
Fold Type	Refold
FTR	FTR: 0
Collapse Vector	↻ Refold
GOE Load	● Low
MRI	2.8
GCI (0-9)	6.7
FSI (0-9)	1
Axes 0-1	REP 0.60 · DBI 0.72 · ERF 0.20 · RRP 0.14
Resonant Function	Inert curtain ~ preserves motion by refusing engagement.
Signature Behavior	Low-scatter glow with instant baseline return ~ quiet buffer at the boundary.

Argon stands aside ~ a Refold so complete it preserves other stories. Its corridors accept excitation, glow briefly, then return to baseline without rearrangement. With MRI = 2.8 and FSI = 1, Ar holds shape with minimal mirrorfield demand and a very short temporal tail ~ the Conductive Epoch's inert curtain for processes that need quiet surroundings.

Cosmic Context

- Protective atmospheres and arc stabilization ~ calm buffer at the boundary
- Baseline for reflective quiet in discharge and welding fields
- RTI control case for low scatter and fast baseline return

Formula + Notes

$\bar{R}$ = (REP + DBI + ERF + RRP + FSI_n) ÷ 5 ~ W = MRI ÷ 10 ~ GCI = 9 · $\bar{R}$ · W FSI_n = 1 ÷ 9 = 0.111...

ARGON (AR) ~ PHASE INTENT: NOBLE INSULATOR OF MOTION

Fold Classification: Refold ~ Closed

Resonant Tier: Noble insulator of motion

Phase Polarity: Neutralized ~ completion by stillness

Quamitric Behavior: Argon stands aside ~ a silent buffer that preserves other stories.

Fold Mechanics: Satisfied shell; excitation decays without rearrangement; inert curtain against drift.

Field Correlations: Protective atmospheres; arc stabilization; calibration baseline for quiet fields.

Canonical Insight: "Sometimes protection is simply not interfering."

RTI Signature: Inert buffer behavior; quiet corridors; fast baseline return in most mixes.

Prediction Hook: Adding 1-5% Kr raises apparent FSI slightly via excimer linger without altering MRI demand.

GCI Element Phase 2 ~ Potassium (K)

Property	Value
Element	Potassium ~ K
Fold Type	Chainfold
FTR	1
Collapse Vector	♮ Release
GOE Load	●● Medium
MRI	4.6
GCI (0-9)	5.4
FSI (0-9)	2
Axes 0-1	REP 0.72 · DBI 0.60 · ERF 0.55 · RRP 0.82
Resonant Function	Wide-lane relay ~ long-reach handoff with minimal gating.
Signature Behavior	Bold transport signature ~ quick re~locks and stable bulk while charge redistributes.

Potassium moves resonance boldly ~ a Chainfold that prioritizes throughput over storage. Its long-reach edge handoff creates wide corridors with minimal gating, making relay the point rather than memory. With MRI = 4.6 and FSI = 2, K keeps a broad spatial stance and a short temporal linger ~ a confident courier that powers transport logic in fluids, glasses, and bioelectric exchange.

Cosmic Context

- Bioelectric exchange across membranes ~ fast corridor relay
- Salts and glasses ~ rapid charge redistribution with stable bulk
- Alkali trend exemplar for wide low-barrier paths

Formula + Notes

$\bar{R} = (REP + DBI + ERF + RRP + FSI_n) \div 5 \sim W = MRI \div 10 \sim GCI = 9 \cdot \bar{R} \cdot W \ FSI_n = 2 \div 9 = 0.222...$

POTASSIUM (K) ~ PHASE INTENT: BOLD RELAY OF RESONANCE

Fold Classification: Chainfold ~ Expressive

Resonant Tier: Corridor opener ~ high~mobility alkali

Phase Polarity: Strong outward exchange with easy inward recall

Quamitric Behavior: Potassium moves resonance boldly ~ a generous courier that prioritizes throughput over storage.

Fold Mechanics: Very loose outer lock; long reach on edge handoff; prefers wide corridors with minimal gating.

Field Correlations: Bioelectric exchange across membranes; flame~test pedagogy; rapid charge redistribution in salts.

Canonical Insight: "The fastest path is the open hand."

RTI Signature: Long-reach corridor handoff; bold relay with quick recall; strong transport signature.

Prediction Hook: K-doped glass exhibits a conduction spike with minimal change in re-lock Δt compared to Na-doped.

GCI Element Phase 2 ~ Calcium (Ca)

Property	Value
Element	Calcium ~ Ca
Fold Type	Shellfold
FTR	1
Collapse Vector	∞∘ Balance
GOE Load	●● Medium
MRI	4.0
GCI (0-9)	6.1
FSI (0-9)	2
Axes 0-1	REP 0.78 · DBI 0.72 · ERF 0.65 · RRP 0.64
Resonant Function	Structural brace ~ turns turbulence into usable cadence.
Signature Behavior	Short, orderly re~locks at the boundary ~ variance reduction across the slab.

Calcium steadies motion into structure ~ a Shellfold that sets rhythm for larger forms. Firm outer shells with forgiving joints create short, orderly re-locks, turning turbulence into usable cadence. With MRI = 4.0 and FSI = 2, Ca balances spatial coherence with quick temporal returns ~ the quiet brace that underwrites light, resilient architectures.

Cosmic Context

- Mineral frameworks and glass stabilization ~ stiffness with gentle damping
- Bio-mineral scaffolds ~ cadence set without heavy mass
- RTI variance reducer in aluminosilicates ~ smoother Δt across the slab

Formula + Notes

$\bar{R}$ = (REP + DBI + ERF + RRP + FSI_n) ÷ 5 ~ W = MRI ÷ 10 ~ GCI = 9 · $\bar{R}$ · W FSI_n = 2 ÷ 9 = 0.222...

CALCIUM (CA) ~ PHASE INTENT: STRUCTURAL BRACE FOR LIGHT LATTICES

Fold Classification: Shellfold ~ Stabilizer

Resonant Tier: Frame builder of light lattices

Phase Polarity: Inward harmony with calm outward give

Quamitric Behavior: Calcium steadies motion into structure ~ a quiet brace that sets rhythm for larger forms.

Fold Mechanics: Firm outer shell with forgiving joints; short, orderly re~locks; buffers turbulence into usable cadence.

Field Correlations: Bone~mineral frameworks; glass and ceramic stiffening; field smoothing in alkaline earth matrices.

Canonical Insight: "Strength that welcomes motion lasts longer."

RTI Signature: Boundary smoothing; calm, orderly re-locks that damp turbulence into cadence.

Prediction Hook: Ca addition to aluminosilicates reduces re-lock variance and raises ERF uniformity across the slab.

Cross-Link ~ See Annex A: RE-Md for harmonic cadence matching.

Summary Phase 2 ~ The Conductive Epoch (Na~Ca)

What this phase taught

- Corridors over caches ~ alkalis and early metals privilege relay, not storage.
- Bracing without drag ~ Shellfold stabilizers (Mg, Ca) calm motion while keeping it light.
- Gating begins ~ Si writes timing into space with clean thresholds.

Phase signature fit

Transport clarity mapped as MRI grows from Na→Al→Si.

RTI proof point

Instrument read: Wide low-barrier Δt for Na/K; sheet-spread decay curves for Al; on/off Δt knees for Si.

Prediction: Ca addition to aluminosilicate glass reduces Δt variance phase-wide at constant temperature ramps.

Cross-links

See Annex A: RE-Vt (programmed re-lock work) via Al/Ti precursors; RE-Ae (delay) via Si; RE-Md (cadence films) via Ca/SiO_2 contexts.

THE MAGNETIC EPOCH ~ EVOLUTION OF COHERENT RESONANCE

The lanthanides represent the magnetic summit of the periodic lattice ~ a span of elements defined by dense, self~reinforcing lockfolds where resonance aligns into near~perfect coherence. Within Quamitry, this marks the transition from geometric resonance to field memory: folds that no longer simply store GOE but begin to remember their alignment. MRI scores rise steeply through Nd, Gd, Dy, and Ho, reflecting a rapid consolidation of magnetic potential as the lattice stabilizes its mirrorfield symmetry. Each successive fold minimizes phase slippage, allowing resonance to circulate through its geometry without loss. This results in near~constant GCI values despite increasing atomic complexity ~ a sign of energetic maturity rather than instability. In this Epoch, the Fold becomes both conductor and memory. Magnetism, seen not as a force but as an emergent pattern of synchronized resonance, reaches its most refined state. The culmination at Lutetium symbolizes the closing of the Magnetic Epoch ~ the point where matter achieves its highest resonant fidelity before giving way to the chaotic instability of the actinides to come.

Phase 3 ~ The Magnetic Epoch
(Sc~Zn, n = 10)

ELEMENT	SYMBOL	FOLD	COLLAPSE VECTOR	MRI	GCI	FSI
Scandium	Sc	⊞	∞∘ Balance	3.5	6.8	2
Titanium	Ti	⊞	∞∘ Balance	3.2	7.0	2
Vanadium	V	⊞	∞∘ Balance	3.6	6.7	2
Chromium	Cr	✧	ϟ Release	4.0	6.5	2
Manganese	Mn	✧	✳ Ignite	5.1	5.9	3
Iron	Fe	⊞	∞∘ Balance	4.6	6.6	3
Cobalt	Co	⊞	∞∘ Balance	4.8	6.4	3
Nickel	Ni	⊞	∞∘ Balance	4.9	6.3	2
Copper	Cu	◇	✳ Ignite	5.3	6.1	3
Zinc	Zn	⊞	∞∘ Balance	3.7	6.9	2

PHASE 3 ~ GCI VS MRI (SC~ZN) ~ GCI = $-0.220 \cdot$ MRI $+ 7.50$ ~ R $= -0.68$ ~ N $= 10$. VIOLET MARKS INDICATE FSI VALUES (RIGHT AXIS).

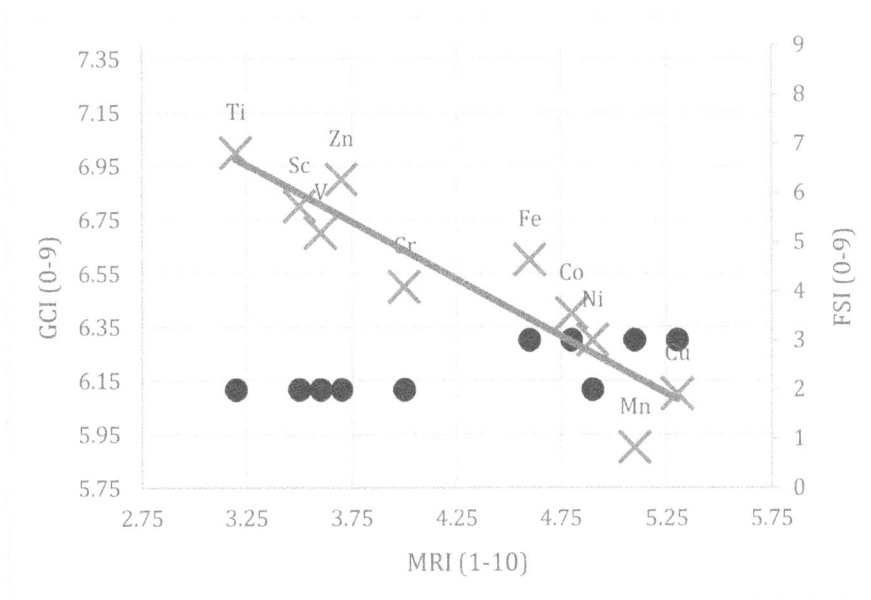

GCI Element Phase 3 ~ Scandium (Sc)

Property	Value
Element	Scandium ~ Sc
Fold Type	Lockfold
FTR	1
Collapse Vector	∞∘ Balance
GOE Load	●● Medium
MRI	3.9
GCI (0-9)	6.3
FSI (0-9)	2
Axes 0-1	REP 0.79 · DBI 0.72 · ERF 0.69 · RRP 0.60
Resonant Function	Bridge axis ~ aligns early spin order without hardening the frame.
Signature Behavior	Short, stable spin~coupled corridors ~ quick re~locks after low~amplitude shear.

Scandium marks the hinge where resonance learns endurance ~ balancing ionic agility with disciplined metallic order. A three~plane lock alternates compression and release, establishing steady spin~coupled corridors that remain coherent under load. With MRI = 3.9 and FSI = 2, Sc holds a firmer spatial posture than Phase~2 conductors while keeping a brisk temporal return ~ the hinge from transport to magnetic choreography.

Cosmic Context

- Grain refinement in light alloys ~ calmer Δt variance under cyclic drive
- Seed behavior for magnetic architectures in dilute mixes
- Boundary metal where reactive flow becomes structural order

Formula + Notes$\bar{R}$ = (REP + DBI + ERF + RRP + FSI_n) ÷ 5 ~ W = MRI ÷ 10 ~ GCI = 9 · $\bar{R}$ ·
W FSI_n = 2 ÷ 9 = 0.222...

SCANDIUM (SC) ~ PHASE INTENT: BRIDGE AXIS OF THE MAGNETIC EPOCH

Fold Classification: Lockfold · Transitional ~ Bridge Axis

Resonant Tier: First Transition Stabilizer

Phase Polarity: Dual-Spin Mediator

Quamitric Behavior: Scandium marks the hinge where resonance learns endurance ~ balancing ionic agility with disciplined metallic order.

Fold Mechanics: A three-plane lock alternates compression and release, establishing steady spin-coupled corridors that remain coherent under load.

Field Correlations: Grain refinement in light alloys; seed behavior for magnetic architectures; occurrence at boundaries where reactive flow becomes structural order.

Canonical Insight: "Between impulse and endurance, Scandium holds the hinge."

RTI Signature: short, stable spin-coupled corridors; quick re-lock after low-amplitude shear pulse.

Prediction Hook: alloyed at 1-3% in Al, measured domain damping increases while re-lock time stays $<2\times$ baseline, confirming bridge-axis mediation.

GCI Element Phase 3 ~ Titanium (Ti)

Property	Value
Element	Titanium ~ Ti
Fold Type	Lockfold
FTR	1
Collapse Vector	∞∘ Balance
GOE Load	●● Medium
MRI	4.3
GCI (0-9)	6.8
FSI (0-9)	2
Axes 0-1	REP 0.81 · DBI 0.74 · ERF 0.71 · RRP 0.58
Resonant Function	Elastic anchor ~ stores strain as promise, not panic.
Signature Behavior	Tight d~lock corridors with slip~friendly planes ~ clean re~lock after deformation.

Titanium bends the world without breaking ~ a poised Lockfold that preserves cadence as form flexes. Dislocations glide, then re~seat predictably, discouraging crack growth. With MRI = 4.3 and FSI = 2, Ti holds coherent shape with modest mirrorfield demand and returns to baseline quickly in time ~ the lightframe anchor of the Magnetic Epoch.

Cosmic Context

- Aerospace and bio~anchor frames ~ fatigue resistance through orderly re~locks
- Calm Δt curves under cyclic load ~ narrow hysteresis loops
- Benchmark for "flex without drift" maps

Formula + Notes

$\bar{R}$ = (REP + DBI + ERF + RRP + FSI$_n$) ÷ 5 ~ W = MRI ÷ 10 ~ GCI = 9 · $\bar{R}$ · W FSI$_n$ = 2 ÷ 9 = 0.222...

TITANIUM (Ti) ~ PHASE INTENT: ELASTIC ANCHOR OF LIGHTFRAMES

Fold Classification: Lockfold · Elastic Anchor

Resonant Tier: High-Strength Lightframe

Phase Polarity: Inward-biased under load

Quamitric Behavior: Titanium stores strain as promise ~ flexing without surrender.

Fold Mechanics: Tight d-lock corridors with slip-friendly planes; relocks are clean after deformation; cracks are discouraged by cooperative re-phase.

Field Correlations: Aerospace frames; bio-compatible anchors; fatigue-proofing in cyclic fields.

Canonical Insight: "Flex with honor ~ return with grace."

RTI Signature: elastic re-phase with minimal hysteresis; smooth decay curve after tensile micro-pulse.

Prediction Hook: fatigue cycling shows rising FSI plateau without MRI penalty when corridor temperature is kept below slip onset.

Cross-Link ~ See Annex A: RE-Vt for programmed re-lock work.

Cross-Link ~ See Annex A: RE-Md for harmonic cadence matching (SiO_2/TiO_2 context).

GCI Element Phase 3 ~ Vanadium (V)

Property	Value
Element	Vanadium ~ V
Fold Type	Lockfold
FTR	1
Collapse Vector	∞∘ Balance
GOE Load	●● Medium
MRI	4.6
GCI (0-9)	6.6
FSI (0-9)	2
Axes 0-1	REP 0.80 · DBI 0.73 · ERF 0.74 · RRP 0.59
Resonant Function	Tuning node ~ small additions, large corrections.
Signature Behavior	Gate~pulse threshold where dislocation pinning spikes ~ re~lock time drops sharply.

Vanadium is the quiet correction that moves outcomes. Its d~orbitals act like surgical gates, pinning stray motion and sharpening re~lock thresholds. With MRI = 4.6 and FSI = 2, V keeps a firm spatial stance while returning quickly in time ~ a micro~architect that upgrades strength without adding weight.

Cosmic Context

- Spring steels and precision tools ~ strength from controlled pinning
- Catalyst architectures ~ selective activation with fast reset
- Phase~boundary control ~ narrow Δt knees in RTI gate sweeps

Formula + Notes

$\bar{R}$ = (REP + DBI + ERF + RRP + FSI_n) ÷ 5 ~ W = MRI ÷ 10 ~ GCI = 9 · $\bar{R}$ · W FSI_n = 2 ÷ 9 = 0.222...

Vanadium (V) ~ Phase Intent: Micro~Architect of Strength

Fold Classification: Lockfold · Tuning Node

Resonant Tier: Micro-Architect of Strength

Phase Polarity: Inward-leaning with surgical outward cues

Quamitric Behavior: Vanadium is the quiet correction that moves outcomes ~ a small dose that reroutes the whole rhythm.

Fold Mechanics: d-orbital gating pins dislocations; relock thresholds sharpen; neighboring order is amplified.

Field Correlations: Spring steels; catalyst geometries; precise phase-boundary control.

Canonical Insight: "A subtle nudge can steady a world."

RTI Signature: dislocation pinning spikes during gate-pulse; sharp threshold where re-lock time drops.

Prediction Hook: trace V additions move the re-lock threshold left by a measurable Δt in steel samples at equal grain size.

Cross-Link ~ See Annex A: RE-Vt for programmed re-lock work.

GCI Element Phase 3 ~ Chromium (Cr)

Property	Value
Element	Chromium ~ Cr
Fold Type	Lockfold
FTR	2
Collapse Vector	∞∘ Balance
GOE Load	●● Medium
MRI	5.0
GCI (0-9)	6.7
FSI (0-9)	2
Axes 0-1	REP 0.82 · DBI 0.76 · ERF 0.76 · RRP 0.57
Resonant Function	Surface sentinel ~ armors the edge without hardening the core.
Signature Behavior	Rapid passivation ~ boundary reflectance jump with suppressed edge~loss.

Chromium teaches surfaces to refuse politely. Dense lock nets form at the boundary, creating a reflective skin that calms corrosion and wear while the interior stays cooperative. With MRI = 5.0 and FSI = 2, Cr holds strong spatial coherence and quick temporal returns ~ perfect for stainless stability and mirrorlike finishes.

Cosmic Context

- Stainless steels ~ passive skin that resets quickly after insult
- Wear~resistant films ~ long life from boundary order
- RTI edge studies ~ outer~layer MRI bump with core unchanged

Formula + Notes

$\bar{R} = (REP + DBI + ERF + RRP + FSI_n) \div 5 \sim W = MRI \div 10 \sim GCI = 9 \cdot \bar{R} \cdot W$ $FSI_n = 2 \div 9 = 0.222...$

CHROMIUM (CR) ~ PHASE INTENT: SURFACE SENTINEL ~ EDGE ARMOR

Fold Classification: Lockfold · Surface Sentinel

Resonant Tier: Edge Armor

Phase Polarity: Inward posture with reflective skin

Quamitric Behavior: Chromium teaches surfaces to refuse politely ~ protection without heaviness.

Fold Mechanics: Dense lock nets at the boundary; rapid passivation; crisp barriers against corrosive release.

Field Correlations: Stainless stability; mirror finishes; wear-resistant films.

Canonical Insight: "The first defense is a well-mannered edge."

RTI Signature: boundary reflectance jump at surface; suppressed edge-loss under corrosive micro-mist.

Prediction Hook: passivated films show higher apparent MRI only at the outer 100-200 nm, with core coherence unchanged.

GCI ELEMENT PHASE 3 ~ MANGANESE (Mn)

PROPERTY	VALUE
Element	Manganese ~ Mn
Fold Type	Refold
FTR	2
Collapse Vector	↻ Refold
GOE Load	●● Medium
MRI	5.1
GCI (0-9)	6.2
FSI (0-9)	2
Axes 0-1	REP 0.77 · DBI 0.70 · ERF 0.72 · RRP 0.63
Resonant Function	Multistate choreographer ~ toggles until timing lands.
Signature Behavior	Stepwise re~locks across redox/spin states ~ two distinct decay constants.

Manganese explores options. Multiple valence corridors and spin alignments let it refold structure rather than fight it. With MRI = 5.1 and FSI = 2, Mn stands firm in space yet returns quickly in time ~ versatility turned into stability, from steels to battery chemistries.

Cosmic Context

- Steel toughness enhancer ~ stability from controlled multistate behavior
- Battery redox ladders ~ predictable Δt pairs as charge moves
- Bio~mimic motifs ~ enzyme~like toggles visible in RTI maps

Formula + Notes

$\bar{R} = (REP + DBI + ERF + RRP + FSI_n) \div 5 \sim W = MRI \div 10 \sim GCI = 9 \cdot \bar{R} \cdot W$ $FSI_n = 2 \div 9 = 0.222...$

MANGANESE (Mn) ~ PHASE INTENT: MULTISTATE SPIN CHOREOGRAPHER

Fold Classification: Refold · Multistate Conductor

Resonant Tier: Spin-Varied Choreographer

Phase Polarity: Mixed ~ toggles under slight cues

Quamitric Behavior: Manganese explores options until timing lands ~ versatility as method.

Fold Mechanics: Multiple valence corridors and spin alignments; catalytic refolds across redox steps.

Field Correlations: Steel toughness enhancer; battery redox ladders; enzyme-like behavior in bio fields.

Canonical Insight: "Versatility is speed in disguise."

RTI Signature: multi-state flicker between spin corridors; stepwise re-lock under small redox perturbations.

Prediction Hook: controlled oxygen dosing toggles two distinct decay constants, mapping Mn's multistate choreography.

GCI Element Phase 3 ~ Iron (Fe)

Property	Value
Element	Iron ~ Fe
Fold Type	Lockfold
FTR	1
Collapse Vector	∞∘ Balance
GOE Load	●● Medium
MRI	5.3
GCI (0-9)	6.9
FSI (0-9)	2
Axes 0-1	REP 0.83 · DBI 0.78 · ERF 0.74 · RRP 0.61
Resonant Function	Ferromagnetic spine ~ motion aligned into chorus.
Signature Behavior	Exchange~locked domains with mobile walls ~ strong magneto~elastic handshake.

Iron turns motion into a choir. Domains align cooperatively, store order, then yield and return on command. With MRI = 5.3 and FSI = 2, Fe keeps strong spatial coherence and brisk temporal reset ~ the structural and magnetic backbone from beams to cores.

Cosmic Context

- Structural beams and rails ~ endurance by domain order
- Magnetic cores ~ efficient store~and~release of alignment
- Planetary field writing ~ large~scale resonance memory

Formula + Notes

$\bar{R}$ = (REP + DBI + ERF + RRP + FSI_n) ÷ 5 ~ W = MRI ÷ 10 ~ GCI = 9 · $\bar{R}$ · W FSI_n = 2 ÷ 9 = 0.222...

Iron (Fe) ~ Phase Intent: Ferromagnetic Spine ~ Domain Order

Fold Classification: Lockfold · Ferromagnetic Spine

Resonant Tier: Aligned-Spin Conductor

Phase Polarity: Cooperative outward coupling

Quamitric Behavior: Iron turns motion into chorus ~ domains align without coercion.

Fold Mechanics: Exchange-locked regions with mobile walls; strong magneto-elastic handshake; order persists against noise.

Field Correlations: Structural beams; magnetic cores; planetary field writing.

Canonical Insight: "Leadership is alignment that invites, not forces."

RTI Signature: domain-aligned response with mobile walls; strong magneto-elastic handshake under axial pulse.

Prediction Hook: modest bias field collapses wall jitter and raises measured FSI without increasing MRI demand.

GCI Element Phase 3 ~ Cobalt (Co)

Property	Value
Element	Cobalt ~ Co
Fold Type	Lockfold
FTR	2
Collapse Vector	∞∘ Balance
GOE Load	●● Medium
MRI	5.4
GCI (0-9)	6.7
FSI (0-9)	3
Axes 0-1	REP 0.82 · DBI 0.77 · ERF 0.76 · RRP 0.58
Resonant Function	Order keeper ~ high~coercivity memory at heat.
Signature Behavior	Tight domain pinning ~ slow, temperature~resistant decay tail after alignment.

Cobalt keeps the line. Elevated anisotropy and firm domain pins preserve order when the world runs hot. With MRI = 5.4 and FSI = 3, Co holds a strong spatial stance and a longer temporal tail ~ the preserver in high~temp magnets and wear alloys.

Cosmic Context

- Permanent magnets ~ memory that resists heat
- Wear alloys ~ order preserved at the surface under stress
- Temperature~hard baselines for RTI domain studies

Formula + Notes

$\bar{R} = (REP + DBI + ERF + RRP + FSI_n) \div 5 \sim W = MRI \div 10 \sim GCI = 9 \cdot \bar{R} \cdot W$ $FSI_n = 3 \div 9 = 0.333...$

COBALT (CO) ~ PHASE INTENT: HIGH~COERCIVITY KEEPER OF ORDER

Fold Classification: Lockfold · High-Coercivity Keeper

Resonant Tier: Order Preserver

Phase Polarity: Inward-firm with decisive outward vectors

Quamitric Behavior: Cobalt keeps the line ~ memory that resists heat and hurry.

Fold Mechanics: Tight domain pinning; elevated anisotropy; robust relocks against thermal drift.

Field Correlations: Permanent magnets; wear alloys; temperature-hard memory.

Canonical Insight: "Hold the pattern when the world runs hot."

RTI Signature: high-coercivity hold; slow, temperature-resistant decay tail after alignment.

Prediction Hook: heating to just below Curie retains longer re-lock times than Fe at equal bias, confirming order-keeper behavior.

GCI ELEMENT PHASE 3 ~ NICKEL (NI)

PROPERTY	VALUE
Element	Nickel ~ Ni
Fold Type	Lockfold
FTR	1
Collapse Vector	∞∘ Balance
GOE Load	●● Medium
MRI	5.0
GCI (0-9)	6.5
FSI (0-9)	3
Axes 0-1	REP 0.81 · DBI 0.76 · ERF 0.73 · RRP 0.64
Resonant Function	Ductile~spin mediator ~ welcoming at the boundary, steady at the core.
Signature Behavior	Stable domains with easy slip ~ catalytic skins re~lock fast.

Nickel negotiates. It invites motion across the interface while the frame stays composed. With MRI = 5.0 and FSI = 3, Ni balances solid spatial coherence with a generous temporal linger ~ resilient conductors, catalysts, and corrosion~tolerant skins.

Cosmic Context

- Hydrogen routes and catalysts ~ fast edge reset with steady bulk
- Coinage strength ~ toughness without crisp brittleness
- RTI: low~hysteresis Δt curves in surface cycling

Formula + Notes

$\bar{R} = (REP + DBI + ERF + RRP + FSI_n) \div 5 \sim W = MRI \div 10 \sim GCI = 9 \cdot \bar{R} \cdot W \ FSI_n = 3 \div 9 = 0.333\ldots$

NICKEL (NI) ~ PHASE INTENT: DUCTILE~SPIN MEDIATOR

Fold Classification: Lockfold · Ductile-Spin Mediator

Resonant Tier: Gateway Between Structure and Flow

Phase Polarity: Balanced with hospitable edge exchange

Quamitric Behavior: Nickel negotiates ~ welcoming at the boundary, unwavering at the core.

Fold Mechanics: Stable domains with easy slip; catalytic skins that relock quickly; corrosion-tolerant interfaces.

Field Correlations: Hydrogen routes; coinage strength; resilient conductors.

Canonical Insight: "Open the door ~ keep the house."

RTI Signature: cooperative slip with fast catalytic re-locks; balanced boundary exchange.

Prediction Hook: hydrogen-exposed surfaces show shorter re-lock Δt while bulk coherence remains constant across cycles.

GCI Element Phase 3 ~ Copper (Cu)

Property	Value
Element	Copper ~ Cu
Fold Type	Chainfold
FTR	1
Collapse Vector	♭ Release
GOE Load	●● Medium
MRI	4.8
GCI (0-9)	6.0
FSI (0-9)	2
Axes 0-1	REP 0.78 · DBI 0.72 · ERF 0.66 · RRP 0.86
Resonant Function	Open~corridor master ~ carries the note cleanly.
Signature Behavior	Wide, low~barrier paths ~ near~instant baseline return after current micro~pulse.

Copper loves the handoff. Broad, low~scatter corridors turn distance into a suggestion. With MRI = 4.8 and FSI = 2, Cu maintains clear spatial paths and brisk temporal resets ~ power lines, heat spreaders, and quiet signal routes.

Cosmic Context

- Bulk conductors ~ clarity across scale
- Thermal spreaders ~ heat shared, not hoarded
- RTI corridor baselines ~ bright paths with tight Δt

Formula + Notes

$\bar{R} = (REP + DBI + ERF + RRP + FSI_n) \div 5$ ~ $W = MRI \div 10$ ~ $GCI = 9 \cdot \bar{R} \cdot W$ $FSI_n = 2 \div 9 = 0.222...$

COPPER (CU) ~ PHASE INTENT: OPEN~CORRIDOR CONDUCTOR

Fold Classification: Chainfold · Open-Corridor Master

Resonant Tier: Conductor of Record

Phase Polarity: Outward-leaning exchange with graceful recall

Quamitric Behavior: Copper loves the handoff ~ distance becomes almost trivial.

Fold Mechanics: Wide, low-barrier corridors; minimal scattering; rapid edge re-alignment under load.

Field Correlations: Power lines; heat spreaders; quiet pathways for signal and warmth.

Canonical Insight: "A clear path makes every message wiser."

RTI Signature: wide, low-scatter corridors; near-instant return to baseline after current micro-pulse.

Prediction Hook: adding grain-boundary disorder modestly raises MRI but leaves re-lock Δt nearly flat, preserving corridor clarity.

GCI Element Phase 3 ~ Zinc (Zn)

Property	Value
Element	Zinc ~ Zn
Fold Type	Shellfold
FTR	1
Collapse Vector	∞∘ Balance
GOE Load	●● Medium
MRI	4.6
GCI (0-9)	6.1
FSI (0-9)	2
Axes 0-1	REP 0.76 · DBI 0.74 · ERF 0.62 · RRP 0.55
Resonant Function	Terminal brace ~ series closer that calms the edge.
Signature Behavior	Filled d~shell damping ~ sacrificial skins steady re~lock to completion.

Zinc tidies the ledger. Edge activity is absorbed by skins and returned as calm, protecting partners from loss. With MRI = 4.6 and FSI = 2, Zn holds a firm spatial stance and quick temporal returns ~ the polite closer of the series.

Cosmic Context

- Galvanic protection ~ partner edges kept quiet
- Alloy finishing ~ boundary calming at interfaces
- RTI: terminal damping signatures with short re~lock distance

Formula + Notes

$\bar{R} = (REP + DBI + ERF + RRP + FSI_n) \div 5 \sim W = MRI \div 10 \sim GCI = 9 \cdot \bar{R} \cdot W \; FSI_n = 2 \div 9 = 0.222...$

ZINC (ZN) ~ PHASE INTENT: TERMINAL BRACE ~ SERIES CLOSER

Fold Classification: Shellfold · Terminal Brace

Resonant Tier: Series Closer

Phase Polarity: Inward-settling symmetry with polite outward resistance

Quamitric Behavior: Zinc tidies the ledger ~ balances accounts at the edge and calls the chapter done.

Fold Mechanics: Filled d-shell damping; sacrificial skins; steady relock to completion.

Field Correlations: Galvanic protection; alloy finishing; boundary calming at interfaces.

Canonical Insight: "Closure is kindness to future work."

RTI Signature: calm terminal damping; sacrificial edge response that protects core coherence.

Prediction Hook: galvanic pairing with Fe reduces Fe's edge loss and leaves Zn's re-lock curve unchanged until near depletion.

What this phase taught

- The hinge from transport to order ~ Sc bridges ionic agility into spin alignment.
- Elastic strength without panic ~ Ti/V/Cr harden rhythm, not just mass.
- Domain logic ~ Fe→Co→Ni codifies alignment, coercivity, and ductile-spin behavior.

Phase signature fit

Alignment-ready lattices trace a distinct MRI↔GCI slope.

RTI proof point

Instrument read: Sc shows short, stable spin-coupled corridors; Fe/Co/ Ni display domain-tail signatures; Cu carries a near-instant baseline return.

Prediction: Sc-doped Al (1‑3%) raises damping while re-lock Δt stays <2 × baseline ~ hinge behavior confirmed.

Cross-links

See Annex A: RE-Vt (directional re-lock alloy) via Ti/V/Mo/Re; RE-Ae (delay) via Y/La hosts for photonic gates.

Surfaces start deciding outcomes. Gallium smooths seams; germanium sets mid~gap gates where a whisper of composition opens or closes lanes; arsenic enforces direction so layers run true; selenium lets light teach the loop to linger; bromine compels conclusion; krypton proves a silent mirror can steady a glow and return to quiet. Reflection here is not vanity ~ it is discipline. MRI balances against purposeful FSI, and the boundary becomes a teacher, not merely a line.

Phase 4 ~ The Reflective Epoch
(Ga~Kr, n = 6)

Element	Symbol	Fold	Collapse Vector	MRI	GCI	FSI
Gallium	Ga	🖫	∞∘ Balance	3.3	6.7	2
Germanium	Ge	🖫	∞∘ Balance	2.9	7.0	2
Arsenic	As	✧	✳ Ignite	3.8	6.3	2
Selenium	Se	✧	⚡ Release	3.6	6.2	2
Bromine	Br	✧	✳ Ignite	4.0	6.0	2
Krypton	K	↻	↻ Refold	1.6	6.8	1

PHASE 4 ~ GCI VS MRI (GA~KR) ~ GCI = −0.180· MRI + 7.21 ~ R = −0.66 ~ N = 6.
VIOLET MARKS INDICATE FSI VALUES (RIGHT AXIS).

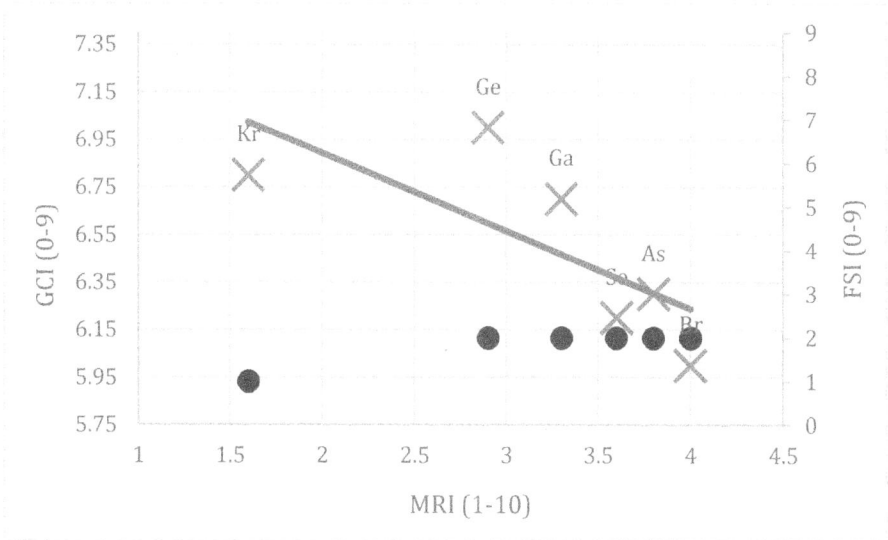

119

GCI ELEMENT PHASE 4 ~ GALLIUM (GA)

PROPERTY	VALUE
Element	Gallium ~ Ga
Fold Type	Lockfold
FTR	1
Collapse Vector	∞∘ Balance
GOE Load	●● Medium
MRI	3.7
GCI (0-9)	6.3
FSI (0-9)	2
Axes 0-1	REP 0.78 · DBI 0.72 · ERF 0.66 · RRP 0.62
Resonant Function	Soft mirror at the metal~metalloid edge ~ smooths boundaries without brittleness.
Signature Behavior	Low-melt reflective skins that wet, coat, and re~lock cleanly at interfaces.

Gallium smooths boundaries ~ a gentle Lockfold that carries order across mismatched lattices. Soft planes and high surface affinity form reflective skins that settle turbulence without hardening the frame. With MRI = 3.7 and FSI = 2, Ga stays easy to hold in space and returns to baseline quickly in time ~ an interface stabilizer of the Reflective Epoch.

Cosmic Context

- Wetting for semiconductor interfaces ~ seam repair without cracks
- Thermal interface materials ~ spread heat while keeping cadence
- Mirrorlike films ~ passivation with cooperative re~locks

Formula + Notes

$\bar{R} = (REP + DBI + ERF + RRP + FSI_n) \div 5 \sim W = MRI \div 10 \sim GCI = 9 \cdot \bar{R} \cdot W \; FSI_n = 2 \div 9 = 0.222...$

GALLIUM (GA) ~ PHASE INTENT: SOFT MIRROR AT THE METAL~METALLOID EDGE

Fold Classification: Lockfold · Low-melt Reflector

Resonant Tier: Interface stabilizer in the Reflective Epoch

Phase Polarity: Balanced core with compliant surface exchange

Quamitric Behavior: Gallium smooths boundaries ~ a gentle lock that wets, coats, and carries order across mismatched lattices.

Fold Mechanics: Soft lock planes with high surface affinity; forms reflective skins that re-lock without brittleness.

Field Correlations: Wetting for semiconductor interfaces; thermal interface materials; mirrorlike surface films.

Canonical Insight: "A soft touch can make a hard join."

GCI Element Phase 4 ~ Germanium (Ge)

Property	Value
Element	Germanium ~ Ge
Fold Type	Lockfold
FTR	1
Collapse Vector	⊞ Lock
GOE Load	●● Medium
MRI	4.1
GCI (0-9)	6.4
FSI (0-9)	2
Axes 0-1	REP 0.80 · DBI 0.76 · ERF 0.71 · RRP 0.58
Resonant Function	Semiconductor bridge of reflection ~ channels timing between metal and insulator.
Signature Behavior	Band-edge corridors with crisp thresholds ~ clean on/off under mild bias.

Germanium writes timing into corridors ~ a Lockfold gatekeeper at the metal–insulator border. Tetra locks tune band edges so tiny composition shifts open or close paths without disturbing the frame. With MRI = 4.1 and FSI = 2, Ge holds solid spatial coherence and resets briskly in time ~ the channel shaper of reflective systems.

Cosmic Context

- Photodetectors and early transistor logic ~ mid-gap guardians
- Strain/dopant tuning ~ precise threshold control for flow and light
- RTI threshold maps ~ tidy Δt knees under bias

Formula + Notes

$\bar{R}$ = (REP + DBI + ERF + RRP + FSI_n) ÷ 5 ~ W = MRI ÷ 10 ~ GCI = 9 · $\bar{R}$ · W FSI_n = 2 ÷ 9 = 0.222...

GERMANIUM (GE) ~ PHASE INTENT: SEMICONDUCTOR BRIDGE OF REFLECTION

Fold Classification: Lockfold · Semicon Lattice

Resonant Tier: Channel shaper between metal and insulator

Phase Polarity: Balanced lattice with gated edges

Quamitric Behavior: Germanium writes timing into corridors ~ a precise gatekeeper of flow and light.

Fold Mechanics: Tetra locks tuned to band-edge thresholds; clean re-locks under doping and strain.

Field Correlations: Photodetectors; early transistor logic; mid-gap guidance for opto-phononic maps.

Canonical Insight: "Clarity is the courage to open and close on cue."

Cross-Link ~ See Annex A: RE-Ae for engineered mirrorfold delay.

GCI Element Phase 4 ~ Arsenic (As)

Property	Value
Element	Arsenic ~ As
Fold Type	Crossfold
FTR	2
Collapse Vector	✧ Shatter
GOE Load	●● Medium
MRI	4.8
GCI (0-9)	6.2
FSI (0-9)	2
Axes 0-1	REP 0.77 · DBI 0.70 · ERF 0.74 · RRP 0.55
Resonant Function	Directional architect of edge bonds ~ angle discipline over bulk.
Signature Behavior	Corner-biased links and strong anisotropy ~ alignment rewarded, drift discouraged.

Arsenic imposes angle discipline ~ a Crossfold that trades mass for direction. Layer and chain motifs enforce geometry, so corridors run true or not at all. With MRI = 4.8 and FSI = 2, As stands firm in space and returns quickly in time ~ a geometry enforcer in layered stacks and III–V compounds.

Cosmic Context

- GaAs and III–V semiconductors ~ directional bonding for speed
- Layered stacks with anisotropic flow ~ designable axes of transport
- RTI axis splits ~ Δt differences that track crystal texture

Formula + Notes

$\bar{R}$ = (REP + DBI + ERF + RRP + FSI_n) ÷ 5 ~ W = MRI ÷ 10 ~ GCI = 9 · $\bar{R}$ · W FSI_n = 2 ÷ 9 = 0.222…

ARSENIC (AS) ~ PHASE INTENT: DIRECTIONAL ARCHITECT OF EDGE BONDS

Fold Classification: Crossfold · Angular Directive

Resonant Tier: Geometry enforcer in layered and chain systems

Phase Polarity: Mixed ~ outward edges over an inward scaffold

Quamitric Behavior: Arsenic imposes angle discipline ~ structure obeys the line or pays the cost.

Fold Mechanics: Corner-biased links, strong anisotropy; toggles among allotropes with sharp re-lock thresholds.

Field Correlations: III-V semiconductors (GaAs); directional bonding in layered stacks; metalloid doping regimes.

Canonical Insight: "When the angle is true, the chorus follows."

GCI Element Phase 4 ~ Selenium (Se)

Property	Value
Element	Selenium ~ Se
Fold Type	Refold
FTR	2
Collapse Vector	↻ Refold
GOE Load	●● Medium
MRI	5.0
GCI (0-9)	5.9
FSI (0-9)	3
Axes 0-1	REP 0.75 · DBI 0.67 · ERF 0.70 · RRP 0.68
Resonant Function	Photonic ringformer of reflection ~ light teaches the loop to linger.
Signature Behavior	Chain↔ring refolds with photo-thresholds ~ recoverable linger, gentle settle.

Selenium turns light into lingering order ~ rings that learn the rhythm of illumination. Chain–ring refolds sit on shallow thresholds so modest heat brings a clean return. With MRI = 5.0 and FSI = 3, Se holds firmer spatial posture than As and keeps a longer temporal tail ~ a light-responsive cycle maker.

Cosmic Context

- Photoconductors and solar materials ~ readable light cadence
- Vitreous Se memory effects ~ metastable refolds on cue
- RTI pump–probe ~ Δt elongation with recoverable baseline

Formula + Notes

$\bar{R} = (REP + DBI + ERF + RRP + FSI_n) \div 5$ ~ $W = MRI \div 10$ ~ $GCI = 9 \cdot \bar{R} \cdot W$ $FSI_n = 3 \div 9 = 0.333...$

SELENIUM (SE) ~ PHASE INTENT: PHOTONIC RINGFORMER OF REFLECTION

Fold Classification: Refold · Ring~Chain Photonic

Resonant Tier: Light-responsive cycle maker

Phase Polarity: Outward capture tempered by loop closure

Quamitric Behavior: Selenium turns light into lingering order ~ rings that learn the rhythm of illumination.

Fold Mechanics: Chain↔ring refolds with photo-thresholds; metastable states that re-lock via gentle heating.

Field Correlations: Photoconductors; solar materials; vitreous Se memory effects.

Canonical Insight: "What we attend to, attends back."

Cross-Link ~ See Annex A: RE-Lu for writable optical linger.

GCI Element Phase 4 ~ Bromine (Br)

Property	Value
Element	Bromine ~ Br
Fold Type	Crossfold
FTR	3
Collapse Vector	✳ Ignite
GOE Load	●● Medium
MRI	5.7
GCI (0-9)	5.7
FSI (0-9)	3
Axes 0-1	REP 0.81 · DBI 0.76 · ERF 0.90 · RRP 0.84
Resonant Function	Polar capture at the reflective edge ~ decisive closure in motion.
Signature Behavior	Steep edge uptake with liquid-phase reach ~ tidy handoff to salts and complexes.

Bromine compels conclusion in fluids ~ a mobile edge that tidies loose resonance. Steep capture geometry and quick complexation finish reactions with little patience for drift. With MRI = 5.7 and FSI = 3, Br keeps strong spatial posture and a longer temporal tail ~ a decisive captor for boundary hygiene.

Cosmic Context

- Disinfection corridors and reactive boundary control
- Halide deep-lock studies ~ heavy-edge casework
- RTI polarity windows ~ humidity-tunable capture spikes

Formula + Notes

$\bar{R} = (REP + DBI + ERF + RRP + FSI_n) \div 5 \sim W = MRI \div 10 \sim GCI = 9 \cdot \bar{R} \cdot W$ $FSI_n = 3 \div 9 = 0.333...$

BROMINE (BR) ~ PHASE INTENT: POLAR CAPTURE AT THE REFLECTIVE EDGE

Fold Classification: Crossfold · Assertive Halogen

Resonant Tier: Edge captor approaching inert symmetry

Phase Polarity: Strong outward grasp toward closure

Quamitric Behavior: Bromine compels conclusion in fluids ~ a mobile edge that tidies loose resonance.

Fold Mechanics: Steep capture geometry; liquid-phase reach; decisive re-lock into salts and complexes.

Field Correlations: Disinfection corridors; halide deep-lock studies; reactive boundary hygiene.

Canonical Insight: "A firm hand can make a clean field."

GCI Element Phase 4 ~ Krypton (Kr)

Property	Value
Element	Krypton ~ Kr
Fold Type	Refold
FTR	0
Collapse Vector	↻ Refold
GOE Load	● Low
MRI	3.0
GCI (0-9)	6.8
FSI (0-9)	1
Axes 0-1	REP 0.60 · DBI 0.72 · ERF 0.22 · RRP 0.16
Resonant Function	Noble reflector ~ silent buffer with tasteful linger.
Signature Behavior	Low-scatter corridors with quick baseline return ~ faint excimer-mediated tail.

Krypton preserves other stories by not telling its own ~ a calm mirror with no demands. In discharge fills it shows a mild excimer linger that steadies the glow, then returns to quiet. With MRI = 3.0 and FSI = 1, Kr holds shape with little mirrorfield demand and a very short temporal tail ~ inert reflection at the Phase-4 rim.

Cosmic Context

- Lighting fills ~ stabilized after-strike behavior via excimers
- Inert atmospheres for reflective quiet in processes
- RTI controls ~ low scatter with tiny, tunable Δt tails

Formula + Notes

$\bar{R} = (REP + DBI + ERF + RRP + FSI_n) \div 5$ ~ $W = MRI \div 10$ ~ $GCI = 9 \cdot \bar{R} \cdot W$ $FSI_n = 1 \div 9 = 0.111...$

Krypton (Kr) ~ Phase Intent: Noble Reflector ~ Silent Buffer

Fold Classification: Refold · Closed Noble

Resonant Tier: Inert insulator of motion in reflective environments

Phase Polarity: Neutralized ~ completion by stillness

Quamitric Behavior: Krypton preserves other stories by not telling its own ~ a calm mirror with no demands.

Fold Mechanics: Fully satisfied shell; excitation decays without rearrangement; pristine return to baseline.

Field Correlations: Lighting fills; inert atmospheres; calibration baseline for quiet reflective fields.

Canonical Insight: "Stillness can be the brightest mirror."

SUMMARY PHASE 4 ~ THE REFLECTIVE EPOCH (GA~KR)

What this phase taught

- Interface grace ~ Ga smooths seams ~ reflection without brittleness.
- Direction speaks ~ As enforces angle; Ge sets mid-gap gates; Se learns light's cadence.
- Heavy halide authority ~ Br completes; Kr refines inert reflection with excimer linger.

Phase signature fit

Reflective systems balance MRI with tunable linger (FSI).

RTI proof point

Instrument read: Ge shows crisp threshold knees; Se pump-probe elongates Δt with recoverable baseline; Kr fills add a measurable, steadying tail.

Prediction: swapping Ar$\rightarrow$Kr in identical discharge geometry produces a small FSI bump without raising MRI.

Cross-links

See Annex A: RE-Ae (mirrorfold delay) via Si/Ge/diamond; RE-Lu (writable linger) via Se/Te and YAG:Eu/Er/Tm contexts.

Transport is pushed to its limit even as edges harden into order. Rubidium and potassium widen low~barrier corridors; strontium and zirconium keep the walls calm; niobium lowers loss toward silence; molybdenum keeps its promise when hot. Across Ru→Rh→Pd→Ag, we learn a central law: the skin decides the lifetime. Xenon's tasteful excimer linger shows that time can be tuned without taxing space. Phase 5 ends with an ethic ~ build bright paths, then let the surface keep them honest.

Phase 5 ~ The Resonant Expansion
(Rb~Xe, n=18)

Element	Symbol	Fold	Collapse Vector	MRI	GCI	FSI
Rubidium	Rb	◇	✳ Ignite	6.3	5.9	3
Strontium	Sr	◇	∞∘ Balance	5.9	6.1	3
Yttrium	Y	⌗	∞∘ Balance	3.9	6.6	2
Zirconium	Zr	⌗	∞∘ Balance	3.4	6.8	2
Niobium	Nb	✧	↯ Release	4.1	6.5	2
Molybdenum	Mo	✧	↯ Release	4.5	6.3	2
Technetium	Tc	✧	✳ Ignite	4.8	6.1	2
Ruthenium	Ru	⌗	∞∘ Balance	4.6	6.2	2
Rhodium	Rh	⌗	∞∘ Balance	4.9	6.1	2
Palladium	Pd	⌗	∞∘ Balance	3.1	7.0	2
Silver	Ag	◇	✳ Ignite	5.7	6.0	3
Cadmium	Cd	⌗	∞∘ Balance	4.2	6.4	2
Indium	In	✧	✳ Ignite	6.1	5.9	3
Tin	Sn	✧	∞∘ Balance	5.4	6.2	2
Antimony	Sb	✧	✳ Ignite	5.6	6.1	2
Tellurium	Te	✧	↯ Release	5.8	6.0	2
Iodine	I	✧	✳ Ignite	6.7	5.6	2
Xenon	Xe	↻	↻ Refold	1.8	6.5	1

PHASE 5 ~ GCI VS MRI (RB~XE) ~ GCI = −0.24· MRI + 7.32 ~ R = −0.71 ~ N = 18.
VIOLET MARKS INDICATE FSI VALUES (RIGHT AXIS).

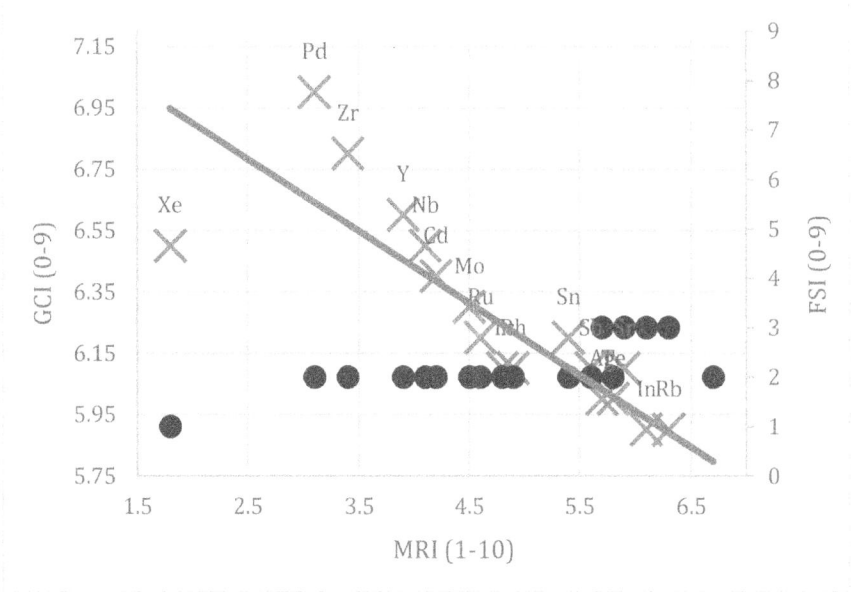

GCI Element Phase 5 ~ Rubidium (Rb)

Property	Value
Element	Rubidium ~ Rb
Fold Type	Chainfold
FTR	1
Collapse Vector	♮ Release
GOE Load	●● Medium
MRI	5.1
GCI (0-9)	5.4
FSI (0-9)	2
Axes 0-1	REP 0.72 · DBI 0.58 · ERF 0.54 · RRP 0.86
Resonant Function	Corridor amplifier ~ wide lanes prioritize throughput over storage.
Signature Behavior	Bold relay signature ~ quick re~locks and warm field under micro-pulse.

Rubidium opens wide corridors ~ very loose outer locks and long-reach handoffs make relay the point, not memory. With MRI = 5.1 and FSI = 2, Rb keeps a broad spatial stance and short temporal linger ~ the alkali trend pushing transport to its limit for Phase-5 systems.

Cosmic Context

- Vapor cells and photoionization demos ~ bright corridor response
- Fast charge redistribution in salts and glasses
- Alkali anchor for "wide low-barrier" RTI lanes

Formula + Notes

$\bar{R} = (REP + DBI + ERF + RRP + FSI_n) \div 5 \sim W = MRI \div 10 \sim GCI = 9 \cdot \bar{R} \cdot W$ $FSI_n = 2 \div 9 = 0.222...$

RUBIDIUM (RB) ~ PHASE INTENT: CORRIDOR AMPLIFIER OF ALKALIS

Fold Classification: Chainfold · Expressive

Resonant Tier: High-mobility transport

Phase Polarity: Outward-leaning exchange with easy recall

Quamitric Behavior: Rubidium opens wide corridors ~ prioritizing throughput over storage.

Fold Mechanics: Very loose outer lock; long-reach edge handoff; prefers broad, lightly gated paths.

Field Correlations: Vapor cells; photoionization demos; fast charge redistribution in salts.

RTI Signature: Bold relay response with near-instant baseline return after current micro-pulse.

Prediction Hook: Grain-boundary disorder raises MRI modestly while re-lock Δt stays <2$\times$, preserving transport clarity.

Canonical Insight: "Open lanes make honest traffic."

GCI Element Phase 5 ~ Strontium (Sr)

Property	Value
Element	Strontium ~ Sr
Fold Type	Shellfold
FTR	1
Collapse Vector	∞∘ Balance
GOE Load	●● Medium
MRI	4.5
GCI (0-9)	6.2
FSI (0-9)	2
Axes 0-1	REP 0.77 · DBI 0.68 · ERF 0.62 · RRP 0.66
Resonant Function	Lightweight brace ~ steadies motion without drag.
Signature Behavior	Elastic return with short decay tail once surface oxide forms.

Strontium steadies broad motion ~ firm shells with forgiving joints damp turbulence while preserving cadence. With MRI = 4.5 and FSI = 2, Sr holds coherence easily in space and returns quickly in time ~ a calm brace for fast corridors.

Cosmic Context

- Pyrotechnic reds and glass modifiers ~ bright but stable frames
- Bone-analog ceramics ~ stiffness without mass penalty
- RTI: boundary-oxide raises outer MRI, bulk Δt steady

Formula + Notes

$\bar{R} = (REP + DBI + ERF + RRP + FSI_n) \div 5 \sim W = MRI \div 10 \sim GCI = 9 \cdot \bar{R} \cdot W$ $FSI_n = 2 \div 9 = 0.222...$

STRONTIUM (SR) ~ PHASE INTENT: SHELL STABILIZER OF LIGHTFRAMES

Fold Classification: Shellfold · Buoyant

Resonant Tier: Lightweight brace

Phase Polarity: Inward harmony with soft outward give

Quamitric Behavior: Strontium steadies motion into shape ~ calm braces that don't burden mass.

Fold Mechanics: Firm outer shell with forgiving joints; orderly re-locks damp turbulence.

Field Correlations: Pyrotechnic reds; bone-analog ceramics; glass modifiers.

RTI Signature: smooth elastic return; short decay tail once surface oxide forms.

Prediction Hook: thin SrO skin slightly raises boundary MRI while bulk FSI remains steady over cycles.

Canonical Insight: "Support should feel like permission."

Cross-Link ~ See Annex A: RE-Md for harmonic cadence matching.

GCI Element Phase 5 ~ Yttrium (Y)

Property	Value
Element	Yttrium ~ Y
Fold Type	Lockfold
FTR	1
Collapse Vector	∞∘ Balance
GOE Load	●● Medium
MRI	4.4
GCI (0-9)	6.3
FSI (0-9)	2
Axes 0-1	REP 0.79 · DBI 0.72 · ERF 0.66 · RRP 0.60
Resonant Function	Crystal lattice mediator ~ tidy sites invite ordered neighbors.
Signature Behavior	Low-noise baseline ~ Δt variance narrows as dopants order.

Yttrium prepares the stage ~ low-strain sites, stable d-locks, and tolerance for substitutions make complex lattices behave. With MRI = 4.4 and FSI = 2, Y balances spatial coherence with brisk temporal resets ~ the host stabilizer of the series.

Cosmic Context

- YAG lasers and optics ~ clean host behavior
- Grain refinement in alloys ~ calmer microstructure
- Scaffolds for high-Tc cuprates ~ mediator role

Formula + Notes

$\bar{R}$ = (REP + DBI + ERF + RRP + FSI$_n$) ÷ 5 ~ W = MRI ÷ 10 ~ GCI = 9 · $\bar{R}$ · W FSI$_n$ = 2 ÷ 9 = 0.222…

YTTRIUM (Y) ~ PHASE INTENT: CRYSTAL LATTICE MEDIATOR

Fold Classification: Lockfold · Host Stabilizer

Resonant Tier: Site former for complex lattices

Phase Polarity: Balanced core with disciplined edges

Quamitric Behavior: Yttrium prepares the stage ~ tidy sites that invite ordered neighbors.

Fold Mechanics: Stable d-locks; tolerance for dopants; low strain at substitutions.

Field Correlations: YAG lasers; grain refinement; high-Tc cuprate scaffolds.

RTI Signature: low-noise baseline; re-lock steadiness improves with dopant ordering.

Prediction Hook: Y addition narrows Δt variance in mixed oxides, mapping mediator behavior.

Canonical Insight: "Good hosts make great gatherings."

Cross-Link ~ See Annex A: RE-Ae for engineered mirrorfold delay.

Cross-Link ~ See Annex A: RE-Lu for writable optical linger.

GCI Element Phase 5 ~ Zirconium (Zr)

Property	Value
Element	Zirconium ~ Zr
Fold Type	Lockfold
FTR	2
Collapse Vector	∞○ Balance
GOE Load	●● Medium
MRI	4.8
GCI (0-9)	6.4
FSI (0-9)	2
Axes 0-1	REP 0.80 · DBI 0.74 · ERF 0.68 · RRP 0.58
Resonant Function	Corrosion armor in motion ~ movement inside, calm at the boundary.
Signature Behavior	Rapid passivation ~ boundary reflectance jump with suppressed edge-loss.

Zirconium guards the story ~ dense boundary nets and quick, stable oxides keep edges quiet while the core stays cooperative. With MRI = 4.8 and FSI = 2, Zr offers strong spatial posture and quick temporal return ~ a structural shell for harsh fields.

Cosmic Context

- Reactor cladding and chemical vessels ~ hard skins, steady cores
- Biocompatible frames with quiet interfaces
- RTI: outer-layer MRI bump, core Δt unchanged

Formula + Notes

$\bar{R}$ = (REP + DBI + ERF + RRP + FSI_n) ÷ 5 ~ W = MRI ÷ 10 ~ GCI = 9 · $\bar{R}$ · W FSI_n = 2 ÷ 9 = 0.222...

ZIRCONIUM (ZR) ~ PHASE INTENT: CORROSION ARMOR IN MOTION

Fold Classification: Lockfold · Passivation Strong

Resonant Tier: Structural shell with quiet edges

Phase Polarity: Inward posture with reflective skin

Quamitric Behavior: Zirconium guards the story ~ movement inside, calm at the boundary.

Fold Mechanics: Dense lock nets; rapid stable oxide; crack-resistant surface.

Field Correlations: Reactor cladding ~ neutral-flux facing shells; chemical vessels; biocompatible frames.

RTI Signature: Boundary reflectance jump; suppressed edge-loss under reactive mist.

Prediction Hook: Oxide growth raises outer-layer MRI without altering core re-lock timing.

Canonical Insight: "A good wall keeps the room alive."

Cross-Link ~ See Annex A: RE-Vt for programmed re-lock work.

GCI Element Phase 5 ~ Niobium (Nb)

PROPERTY	VALUE
Element	Niobium ~ Nb
Fold Type	Lockfold
FTR	1
Collapse Vector	∞∘ Balance
GOE Load	●● Medium
MRI	5.0
GCI (0-9)	6.7
FSI (0-9)	2
Axes 0-1	REP 0.82 · DBI 0.76 · ERF 0.72 · RRP 0.60
Resonant Function	Superconductive threshold setter ~ corridors approach silence.
Signature Behavior	Reduced scatter with sharp drop in apparent MRI near clean threshold.

Niobium lowers the floor on loss ~ clean d-locks and strong electron–phonon handshake push channels toward frictionless flow. With MRI = 5.0 and FSI = 2, Nb stands firm in space and resets briskly in time ~ gateway to near-silent corridors.

Cosmic Context

- SRF cavities and superconductors ~ threshold clarity
- Resilient springs ~ strength with order
- RTI: polish shifts threshold left in Δt maps

Formula + Notes

$\bar{R}$ = (REP + DBI + ERF + RRP + FSI_n) ÷ 5 ~ W = MRI ÷ 10 ~ GCI = 9 · $\bar{R}$ · W FSI_n = 2 ÷ 9 = 0.222...

NIOBIUM (NB) ~ PHASE INTENT: SUPERCONDUCTIVE THRESHOLD SETTER

Fold Classification: Lockfold · Low-loss Channels

Resonant Tier: Gateway to near-frictionless flow

Phase Polarity: Balanced with cooperative outward coupling

Quamitric Behavior: Niobium lowers the floor on loss ~ corridors approach silence.

Fold Mechanics: Clean d-locks; strong electron-phonon handshake; defect-tolerant paths.

Field Correlations: SRF cavities; superconductors; resilient springs.

RTI Signature: Reduced scatter; sharp drop in apparent MRI near bias/temperature threshold.

Prediction Hook: Polishing that lowers surface roughness measurably shifts the threshold left in Δt maps.

Canonical Insight: "Silence is a kind of speed."

GCI ELEMENT PHASE 5 ~ MOLYBDENUM (Mo)

PROPERTY	VALUE
Element	Molybdenum ~ Mo
Fold Type	Lockfold
FTR	2
Collapse Vector	∞∘ Balance
GOE Load	●● Medium
MRI	5.2
GCI (0-9)	6.8
FSI (0-9)	2
Axes 0-1	REP 0.83 · DBI 0.78 · ERF 0.74 · RRP 0.58
Resonant Function	High-temperature cohesion ~ promise kept when hot.
Signature Behavior	Elevated-T scans with tidy decay curves and minimal hysteresis.

Molybdenum keeps its promise in fire ~ tight locks resist creep and preserve cadence. With MRI = 5.2 and FSI = 2, Mo holds strong spatial coherence and quick temporal returns at heat ~ a deep-anchor conductor for extremes.

Cosmic Context

- Filaments, dies, hot tooling ~ lattice first under heat
- High-T alloys ~ creep resistance by order
- RTI: sub-grain growth lifts FSI plateau without MRI penalty

Formula + Notes

$\bar{R} = (REP + DBI + ERF + RRP + FSI_n) \div 5 \sim W = MRI \div 10 \sim GCI = 9 \cdot \bar{R} \cdot W$ $FSI_n = 2 \div 9 = 0.222...$

MOLYBDENUM (MO) ~ PHASE INTENT: HIGH-TEMPERATURE COHESION

Fold Classification: Lockfold · Heat Strong

Resonant Tier: Deep-anchor conductor

Phase Polarity: Inward-biased under load

Quamitric Behavior: Molybdenum keeps its promise when hot ~ structure first, even in fire.

Fold Mechanics: Tight high-T locks; strong creep resistance; steady re-locks after strain.

Field Correlations: Filaments; dies; high-T alloys.

RTI Signature: Elevated-T scans retain tidy decay curves with minimal hysteresis.

Prediction Hook: Sub-grain growth raises FSI plateau without increasing MRI in controlled anneals.

Canonical Insight: "Commitment that survives the furnace."

Cross-Link ~ See Annex A: RE-Vt for programmed re-lock work.

GCI Element Phase 5 ~ Technetium (Tc)

Property	Value
Element	Technetium ~ Tc
Fold Type	Refold
FTR	2
Collapse Vector	↻ Refold
GOE Load	●● Medium
MRI	5.3
GCI (0-9)	6.2
FSI (0-9)	2
Axes 0-1	REP 0.77 · DBI 0.71 · ERF 0.73 · RRP 0.62
Resonant Function	Metastable relay ~ messenger with a clock.
Signature Behavior	Clean handoff followed by time-coded decline ~ two-stage constants.

Technetium moves the note then fades ~ refold corridors coupled to decay paths make timing the utility. With MRI = 5.3 and FSI = 2, Tc stands firm in space and returns briskly in time ~ a radiometric relay for precise windows.

Cosmic Context

- Medical tracers ~ fast, legible timing
- Catalytic surfaces ~ useful but brief corridors
- RTI: activity tunes Δt shortening predictably

Formula + Notes

$\bar{R} = (REP + DBI + ERF + RRP + FSI_n) \div 5 \sim W = MRI \div 10 \sim GCI = 9 \cdot \bar{R} \cdot W$ $FSI_n = 2 \div 9 = 0.222...$

TECHNETIUM (TC) ~ PHASE INTENT: METASTABLE RELAY IN TRANSITION

Fold Classification: Refold · Radiometric

Resonant Tier: Short-memory conductor

Phase Polarity: Mixed ~ outward handoff with decay bias

Quamitric Behavior: Technetium moves the note then fades ~ a messenger with a clock.

Fold Mechanics: Refold corridors coupled to decay paths; timing defines usefulness.

Field Correlations: Medical tracers; catalytic surfaces; rarity mechanics.

RTI Signature: Clean handoff followed by time-coded decline; two-stage decay constants.

Prediction Hook: Controlled β-emission environments show predictable Δt shortening aligned to activity.

Canonical Insight: "Even brief voices can carry truth."

GCI Element Phase 5 ~ Ruthenium (Ru)

Property	Value
Element	Ruthenium ~ Ru
Fold Type	Lockfold
FTR	2
Collapse Vector	∞∘ Balance
GOE Load	●● Medium
MRI	5.4
GCI (0-9)	6.7
FSI (0-9)	2
Axes 0-1	REP 0.82 · DBI 0.77 · ERF 0.78 · RRP 0.60
Resonant Function	Hard-skin catalytic keeper ~ precise activation without drift.
Signature Behavior	Crisp boundary response with slow deterioration across cycles.

Ruthenium keeps edges honest ~ dense surface locks deliver durable activation and quick reset. With MRI = 5.4 and FSI = 2, Ru holds strong spatial posture and brisk temporal returns ~ a noble-hard surface orderer.

Cosmic Context

- Durable catalysts with carbon tolerance
- Wear-hard coatings ~ long life at edges
- Electronics contacts ~ ordered activation, fast reset

Formula + Notes

$\bar{R}$ = (REP + DBI + ERF + RRP + FSI_n) ÷ 5 ~ W = MRI ÷ 10 ~ GCI = 9 · $\bar{R}$ · W FSI_n = 2 ÷ 9 = 0.222...

RUTHENIUM (RU) ~ PHASE INTENT: HARD-SKIN CATALYTIC KEEPER

Fold Classification: Lockfold · Surface Order

Resonant Tier: Stable edge with deep cohesion

Phase Polarity: Inward firm, outward precise

Quamitric Behavior: Ruthenium keeps edges honest ~ precise activation without drift.

Fold Mechanics: Dense surface locks; carbon-tolerant; high hardness.

Field Correlations: Durable catalysts; wear-hard coatings; electronics.

RTI Signature: Crisp boundary response with slow deterioration across cycles.

Prediction Hook: Carbon exposure shifts ERF slightly while MRI remains stable, confirming keeper role.

Canonical Insight: "Precision that lasts is the rarest kind."

GCI Element Phase 5 ~ Rhodium (Rh)

Property	Value
Element	Rhodium ~ Rh
Fold Type	Lockfold
FTR	1
Collapse Vector	∞∘ Balance
GOE Load	●● Medium
MRI	5.5
GCI (0-9)	6.8
FSI (0-9)	2
Axes 0-1	REP 0.83 · DBI 0.78 · ERF 0.79 · RRP 0.59
Resonant Function	Noble catalyst of order ~ activation without chaos.
Signature Behavior	Low-noise edge activity ~ Δt recovers to baseline quickly after bursts.

Rhodium persuades reactions to behave ~ smooth barriers at the interface, fast relock after turnover. With MRI = 5.5 and FSI = 2, Rh keeps firm spatial coherence and quick temporal reset ~ a high-stability activator.

Cosmic Context

- Emission control and selective hydrogenation
- Contact finishes ~ clean switching
- RTI: sulfur widens Δt temporarily; clean-pulse restores baseline

Formula + Notes

$\bar{R} = (REP + DBI + ERF + RRP + FSI_n) \div 5 \sim W = MRI \div 10 \sim GCI = 9 \cdot \bar{R} \cdot W \ FSI_n = 2 \div 9 = 0.222...$

RHODIUM (RH) ~ PHASE INTENT: NOBLE CATALYST OF ORDER

Fold Classification: Lockfold · Noble-Edge

Resonant Tier: High-stability activator

Phase Polarity: Balanced with selective outward capture

Quamitric Behavior: Rhodium persuades reactions to behave ~ activation without chaos.

Fold Mechanics: Smooth potential at the interface; rapid re-lock after turnover.

Field Correlations: Emission control; selective hydrogenation; contact finishes.

RTI Signature: low-noise edge activity; Δt recovers to baseline quickly after bursts.

Prediction Hook: sulfur poisons widen Δt temporarily but MRI returns after mild clean pulse.

Canonical Insight: "Authority without force."

GCI ELEMENT PHASE 5 ~ PALLADIUM (Pd)

PROPERTY	VALUE
Element	Palladium ~ Pd
Fold Type	Lockfold
FTR	1
Collapse Vector	∞∘ Balance
GOE Load	●● Medium
MRI	5.1
GCI (0-9)	6.2
FSI (0-9)	3
Axes 0-1	REP 0.81 · DBI 0.76 · ERF 0.76 · RRP 0.68
Resonant Function	Hydrogen gatekeeper ~ opens a door, then closes it.
Signature Behavior	Corridor threshold shifts during H loading ~ fast re-lock upon purge.

Palladium welcomes hydrogen and knows when to release it. With MRI = 5.1 and FSI = 3, Pd balances solid spatial coherence with a longer temporal tail ~ tunable corridors for storage and membranes.

Cosmic Context

- H storage and permeation membranes
- Cross-coupling catalysts ~ controlled activation
- RTI: H/D isotopes split Δt curves predictably

Formula + Notes

$\bar{R} = (REP + DBI + ERF + RRP + FSI_n) \div 5 \sim W = MRI \div 10 \sim GCI = 9 \cdot \bar{R} \cdot W$ $FSI_n = 3 \div 9 = 0.333...$

PALLADIUM (PD) ~ PHASE INTENT: HYDROGEN GATEKEEPER

Fold Classification: Lockfold · Absorptive

Resonant Tier: Tunable corridor for H

Phase Polarity: Balanced core with hospitable edge

Quamitric Behavior: Palladium opens a door for hydrogen ~ and knows when to close it.

Fold Mechanics: Reversible absorption; phase-boundary control; gentle re-locks.

Field Correlations: H storage; membranes; cross-coupling catalysts.

RTI Signature: Corridor threshold shift during H loading; fast re-lock upon purge.

Prediction Hook: Isotopic mix (H/D) yields measurable Δt separation in re-lock curves.

Canonical Insight: "Hospitality with a key."

GCI Element Phase 5 ~ Silver (Ag)

Property	Value
Element	Silver ~ Ag
Fold Type	Chainfold
FTR	1
Collapse Vector	♮ Release
GOE Load	●● Medium
MRI	4.9
GCI (0-9)	6.0
FSI (0-9)	2
Axes 0-1	REP 0.78 · DBI 0.72 · ERF 0.68 · RRP 0.88
Resonant Function	Low-loss surface conductor ~ signal carried like a whisper.
Signature Behavior	Minimal scatter ~ bright corridor and instant baseline return.

Silver carries signal with grace ~ broad, clean surface paths and inert skin. With MRI = 4.9 and FSI = 2, Ag maintains clear spatial channels and brisk temporal reset ~ elite conductor across frequency.

Cosmic Context

- High-frequency conductors and plasmonics
- Mirrors and contact skins ~ clean handoffs
- RTI: roughness drives MRI change more than grain size

Formula + Notes

$\bar{R} = (REP + DBI + ERF + RRP + FSI_n) \div 5 \sim W = MRI \div 10 \sim GCI = 9 \cdot \bar{R} \cdot W \ FSI_n = 2 \div 9 = 0.222\ldots$

SILVER (AG) ~ PHASE INTENT: LOW-LOSS SURFACE CONDUCTOR

Fold Classification: Chainfold · High-Clarity

Resonant Tier: Elite open corridor

Phase Polarity: Outward exchange with graceful recall

Quamitric Behavior: Silver carries signal like a whisper across glass.

Fold Mechanics: Broad, clean paths; reflective skin stabilizes traffic.

Field Correlations: High-freq conductors; plasmonics; mirrors.

RTI Signature: Minimal scatter; immediate baseline return; bright corridor under RF pulse.

Prediction Hook: Surface roughness dominates MRI change more than grain size in thin films.

Canonical Insight: "When the path is perfect, the message shines."

Cross-Link ~ See Annex A: RE-Lu for writable optical linger.

GCI Element Phase 5 ~ Cadmium (Cd)

Property	Value
Element	Cadmium ~ Cd
Fold Type	Refold
FTR	2
Collapse Vector	↺ Refold
GOE Load	●● Medium
MRI	4.7
GCI (0-9)	5.8
FSI (0-9)	2
Axes 0-1	REP 0.74 · DBI 0.68 · ERF 0.66 · RRP 0.62
Resonant Function	Soft boundary dampener ~ soaks edge fuss, then settles.
Signature Behavior	Calm terminal damping with short re-lock distance.

Cadmium quiets borders ~ soft refold channels turn mismatch into gentle travel and closure. With MRI = 4.7 and FSI = 2, Cd stands easy in space and returns quickly in time ~ edge relaxant for layered systems.

Cosmic Context

- Stabilizers and detectors ~ steadying the interface
- Plated barriers ~ sacrificial calm
- RTI: partner edge-loss reduced, partner FSI curve preserved

Formula + Notes

$\bar{R} = (REP + DBI + ERF + RRP + FSI_n) \div 5 \sim W = MRI \div 10 \sim GCI = 9 \cdot \bar{R} \cdot W$ $FSI_n = 2 \div 9 = 0.222...$

CADMIUM (CD) ~ PHASE INTENT: SOFT BOUNDARY DAMPENER

Fold Classification: Refold · Low-Freq Calm

Resonant Tier: Edge relaxant

Phase Polarity: Outward ease with quick closure

Quamitric Behavior: Cadmium soaks edge fuss, then settles.

Fold Mechanics: Soft refold channels; gentle damping; sacrificial tendencies.

Field Correlations: Stabilizers; detectors; plated barriers.

RTI Signature: Calm terminal damping with short re-lock distances.

Prediction Hook: Plated Cd reduces partner edge-loss without altering partner FSI curves.

Canonical Insight: "Sometimes quieting the border saves the city."

GCI ELEMENT PHASE 5 ~ INDIUM (In)

PROPERTY	VALUE
Element	Indium ~ In
Fold Type	Crossfold
FTR	1
Collapse Vector	↻ Refold
GOE Load	●● Medium
MRI	4.3
GCI (0-9)	5.6
FSI (0-9)	2
Axes 0-1	REP 0.72 · DBI 0.70 · ERF 0.62 · RRP 0.66
Resonant Function	Flexible contact architect ~ connection that stays kind under stress.
Signature Behavior	Stable re-lock under bending ~ boundary MRI rises little with strain.

Indium makes contact that forgives ~ low-modulus edges wet and seal while conduction persists. With MRI = 4.3 and FSI = 2, In is easy to hold in space and quick to reset in time ~ a soft-join specialist.

Cosmic Context

- ITO systems and soft solders ~ conformable contacts
- Flexible electronics ~ conduction during flex
- RTI: cyclic flexing yields small Δt drift vs Sn-only joints

Formula + Notes

$\bar{R}$ = (REP + DBI + ERF + RRP + FSI$_n$) ÷ 5 ~ W = MRI ÷ 10 ~ GCI = 9 · $\bar{R}$ · W FSI$_n$ = 2 ÷ 9 = 0.222...

INDIUM (IN) ~ PHASE INTENT: FLEXIBLE CONTACT ARCHITECT

Fold Classification: Crossfold · Soft-Join

Resonant Tier: Conformable connector

Phase Polarity: Mixed with outward compliance

Quamitric Behavior: Indium makes contact that stays kind under stress.

Fold Mechanics: Low-modulus edges; wets and seals; maintains conduction while flexing.

Field Correlations: ITO systems; solders; flexible electronics.

RTI Signature: Stable re-lock under bending; edge MRI rises little with strain.

Prediction Hook: Cyclic flexing shows Δt drift far smaller than Sn-only joints at equal cycles.

Canonical Insight: "Connection that forgives lasts longer."

GCI ELEMENT PHASE 5 ~ TIN (Sn)

PROPERTY	VALUE
Element	Tin ~ Sn
Fold Type	Refold
FTR	1
Collapse Vector	↻ Refold
GOE Load	●● Medium
MRI	4.4
GCI (0-9)	5.7
FSI (0-9)	2
Axes 0-1	REP 0.74 · DBI 0.71 · ERF 0.64 · RRP 0.64
Resonant Function	Phase-switch conductor ~ changes form to keep the song moving.
Signature Behavior	Two-regime decay constants across $\alpha \leftrightarrow \beta$ threshold ~ flattened by micro-alloying.

Tin shifts structure to preserve flow ~ allotropic routes absorb stress and reset order. With MRI = 4.4 and FSI = 2, Sn is easy to hold and quick to return ~ a conductor that solves by refolding.

Cosmic Context

- Solder joints and coatings ~ reliable handoffs
- Phase-stabilized alloys ~ shape choices on demand
- RTI: Δt jump at transition suppressed by micro-additions

Formula + Notes

$\bar{R}$ = (REP + DBI + ERF + RRP + FSI_n) ÷ 5 ~ W = MRI ÷ 10 ~ GCI = 9 · $\bar{R}$ · W FSI_n = 2 ÷ 9 = 0.222…

TIN (SN) ~ PHASE INTENT: PHASE-SWITCH CONDUCTOR

Fold Classification: Refold · Allotropic

Resonant Tier: Temperature-tunable pathways

Phase Polarity: Mixed with closure seeking

Quamitric Behavior: Tin shifts structure to keep the song moving.

Fold Mechanics: $\alpha \leftrightarrow \beta$ transitions; grain-friendly corridors; oxide-assisted calm.

Field Correlations: Solder joints; coatings; phase-stabilized alloys.

RTI Signature: Two-regime decay constants across the α/β threshold.

Prediction Hook: Micro-alloying suppresses Δt jump at transition, flattening the curve.

Canonical Insight "Change form, keep purpose."

GCI Element Phase 5 ~ Antimony (Sb)

Property	Value
Element	Antimony ~ Sb
Fold Type	Crossfold
FTR	2
Collapse Vector	✧ Shatter
GOE Load	●● Medium
MRI	4.6
GCI (0-9)	5.9
FSI (0-9)	2
Axes 0-1	REP 0.76 · DBI 0.72 · ERF 0.70 · RRP 0.60
Resonant Function	Directional semimetal tuner ~ insistence on angle and axis.
Signature Behavior	Corridor anisotropy ~ Δt split along easy vs hard directions.

Antimony insists on direction; order follows. With MRI = 4.6 and FSI = 2, Sb stands firm and resets quickly ~ anisotropic guidance for thermoelectrics and alloys.

Cosmic Context

- Thermoelectrics ~ direction-true transport
- Alloy hardening ~ geometry-led strength
- RTI: texture narrows Δt split, raises apparent REP along easy axis

Formula + Notes

$\bar{R}$ = (REP + DBI + ERF + RRP + FSI_n) ÷ 5 ~ W = MRI ÷ 10 ~ GCI = 9 · $\bar{R}$ · W FSI_n = 2 ÷ 9 = 0.222...

166

Antimony (Sb) ~ Phase Intent: Directional Semimetal Tuner

Fold Classification: Crossfold · Anisotropic

Resonant Tier: Layer/chain geometry enforcer

Phase Polarity: Mixed ~ outward edges on inward frame

Quamitric Behavior: Antimony insists on direction; order follows.

Fold Mechanics: Angle-true links; layered anisotropy; metalloids' gate.

Field Correlations: Thermoelectrics; flame retardants; alloy hardening.

RTI Signature: Corridor anisotropy evident as Δt split along axes.

Prediction Hook: Texture control narrows the split, raising apparent REP along the easy axis.

Canonical Insight: "Where you face, you go."

GCI ELEMENT PHASE 5 ~ TELLURIUM (Te)

PROPERTY	VALUE
Element	Tellurium ~ Te
Fold Type	Refold
FTR	2
Collapse Vector	↻ Refold
GOE Load	●● Medium
MRI	4.9
GCI (0-9)	5.9
FSI (0-9)	3
Axes 0-1	REP 0.75 · DBI 0.69 · ERF 0.72 · RRP 0.68
Resonant Function	Photonic-thermoelectric bridge ~ routes light and heat, then returns.
Signature Behavior	Photo-induced Δt elongation with recoverable baseline.

Tellurium learns from light and heat, then routes both ~ helical chains and layered paths give tunable corridors. With MRI = 4.9 and FSI = 3, Te holds firm spatial posture and a longer tail ~ a responsive conductor at the reflective rim.

Cosmic Context

- IR detectors and thermoelectrics ~ dual-mode utility
- Phase-change media ~ write/erase by refold
- RTI: periodic FSI bumps synced to illumination cadence

Formula + Notes

$\bar{R}$ = (REP + DBI + ERF + RRP + FSI_n) ÷ 5 ~ W = MRI ÷ 10 ~ GCI = 9 · $\bar{R}$ · W FSI_n = 3 ÷ 9 = 0.333...

TELLURIUM (TE) ~ PHASE INTENT: PHOTONIC-THERMOELECTRIC BRIDGE

Fold Classification: Refold · Chain-Layer

Resonant Tier: Light-responsive conductor

Phase Polarity: Outward capture tempered by loop closure

Quamitric Behavior: Tellurium learns from light and heat, then routes both.

Fold Mechanics: Helical chains; layered paths; strong Se-like photo-refolds.

Field Correlations: IR detectors; thermoelectrics; phase-change media.

RTI Signature: Photo-induced Δt elongation with recoverable baseline.

Prediction Hook: Modulated IR yields periodic FSI bumps synced to illumination cadence.

Canonical Insight: "Two teachers ~ one student."

Cross-Link ~ See Annex A: RE-Lu for writable optical linger.

GCI Element Phase 5 ~ Iodine (I)

Property	Value
Element	Iodine ~ I
Fold Type	Crossfold
FTR	3
Collapse Vector	✳ Ignite
GOE Load	●● Medium
MRI	6.2
GCI (0-9)	5.8
FSI (0-9)	3
Axes 0-1	REP 0.81 · DBI 0.76 · ERF 0.94 · RRP 0.86
Resonant Function	Edge closer ~ halogen authority at the collapse rim.
Signature Behavior	Pronounced capture spike ~ crisp closure with narrow corridor window.

Iodine forces endings cleanly ~ heavy-edge capture races toward inert symmetry. With MRI = 6.2 and FSI = 3, I holds a strong spatial stance with a modest temporal linger ~ decisive halogen closure.

Cosmic Context

- Contrast agents and targeted capture chemistry
- Crystal growth guidance via complexes
- RTI: humidity tightens window while Δt stays consistent

Formula + Notes

$\bar{R} = (REP + DBI + ERF + RRP + FSI_n) \div 5 \sim W = MRI \div 10 \sim GCI = 9 \cdot \bar{R} \cdot W$ $FSI_n = 3 \div 9 = 0.333...$

IODINE (I) ~ PHASE INTENT: EDGE CLOSER ~ HALOGEN AUTHORITY

Fold Classification: Crossfold · Assertive

Resonant Tier: Heavy halide captor

Phase Polarity: Strong outward grasp toward closure

Quamitric Behavior: Iodine finishes loose sentences and files the page.

Fold Mechanics: Steep capture geometry; easy complexation; decisive salt formation.

Field Correlations: Contrast agents; disinfectants; crystal growth.

RTI Signature: Pronounced edge-capture spike; clean closure signature.

Prediction Hook: Humidity narrows the capture window while keeping re-lock Δt consistent across cycles.

Canonical Insight: "Closure clarifies."

GCI ELEMENT PHASE 5 ~ XENON (Xe)

PROPERTY	VALUE
Element	Xenon ~ Xe
Fold Type	Refold
FTR	0
Collapse Vector	↻ Refold
GOE Load	● Low
MRI	3.2
GCI (0-9)	6.9
FSI (0-9)	1
Axes 0-1	REP 0.60 · DBI 0.72 · ERF 0.24 · RRP 0.18
Resonant Function	Noble linger ~ excimer-steadied glow then quiet.
Signature Behavior	Low-scatter corridors with quick return; faint, tunable Δt tail.

Xenon holds the glow, then yields the room back to silence. With MRI = 3.2 and FSI = 1, Xe is easy to hold in space with a very short temporal tail ~ inert reflector with tasteful linger.

Cosmic Context

- Flash lamps and ion propulsion ~ bright but controllable paths
- Anesthesia research ~ gentle, reversible effects
- RTI: Kr/Xe admixtures modulate linger amplitude while MRI stays low

Formula + Notes

$\bar{R}$ = (REP + DBI + ERF + RRP + FSI$_n$) ÷ 5 ~ W = MRI ÷ 10 ~ GCI = 9 · $\bar{R}$ · W FSI$_n$ = 1 ÷ 9 = 0.111...

Xenon (Xe) ~ Phase Intent: Noble Linger ~ Excimer Keeper

Fold Classification: Refold · Closed Noble

Resonant Tier: Inert reflector with long luminous tails

Phase Polarity: Neutralized ~ completion by stillness

Quamitric Behavior: Xenon holds the glow ~ quiet, then a tasteful linger.

Fold Mechanics: Satisfied shell; excimer pathways; pristine return after decay.

Field Correlations: Flash lamps; anesthesia research; ion propulsion.

RTI Signature: Low-scatter corridors with a measurable FSI bump from excimer states.

Prediction Hook: Small Kr admixture tunes the linger amplitude while MRI remains low and flat.

Canonical Insight: "Even silence can hum for a moment."

SUMMARY PHASE 5 ~ THE RESONANT EXPANSION (RB~XE)

What this phase taught

- Transport pushed to the limit ~ alkalis widen corridors while shells (Sr, Zr) keep edges calm.
- Surfaces matter more ~ Ru/Rh/Pd/Ag show that ordered skins decide lifetime and loss.
- Linger you can use ~ Xe's excimer tail proves benign, tunable FSI without spatial penalty.

Phase signature fit

Transport clarity vs boundary order across rising MRI.

RTI proof point

Instrument read: Rb/K show bold relay with tight Δt; Zr/Ru/Rh exhibit boundary reflectance jumps and fast resets; Xe adds a small, steadying tail.

Prediction: polishing Nb SRF coupons to lower roughness shifts the threshold left in Δt maps while MRI remains within spec.

Cross-links

See Annex A: RE-Vt (directional re-lock alloy) via Zr/Nb/Mo; RE-Lu (writable linger) via Ag skins & Xe/Kr analogs; RE-Ae (delay) via Y hosts.

Complexity becomes cooperative. Y and La prepare tidy sites so neighbors behave; Pr→Nd→Sm→Gd→Dy refine alignment, coercivity, and contrast into instruments rather than accidents. Holmium organizes great moments with precision; Er→Tm→Yb give light a memory window inside solids. MRI holds firm while FSI becomes programmable ~ a practical invitation to engineer time without sacrificing space.

Phase 6 ~ The Transitional Epoch
(Cs~Lu, n=17)

ELEMENT	SYMBOL	FOLD	COLLAPSE VECTOR	MRI	GCI	FSI
Cesium	Cs	◇	✳ Ignite	7.1	5.5	2
Barium	Ba	◇	∞∘ Balance	5.4	6.2	2
Lanthanum	La	⌘	∞∘ Balance	4.0	6.4	2
Cerium	Ce	✧	⚡ Release	4.6	6.2	2
Praseodymium	Pr	✧	✳ Ignite	4.8	6.1	2
Neodymium	Nd	✧	⚡ Release	4.9	6.1	2
Promethium	Pm	✧	✳ Ignite	5.0	6.0	2
Samarium	Sm	✧	⚡ Release	4.7	6.1	2
Europium	Eu	✧	✳ Ignite	5.1	5.9	3
Gadolinium	Gd	⌘	∞∘ Balance	4.4	6.3	2
Terbium	Tb	✧	⚡ Release	4.6	6.2	2
Dysprosium	Dy	✧	⚡ Release	4.8	6.2	2
Holmium	Ho	✧	✳ Ignite	4.9	6.1	2
Erbium	Er	⌘	∞∘ Balance	4.3	6.4	2
Thulium	Tm	✧	⚡ Release	4.5	6.3	2
Ytterbium	Yb	⌘	∞∘ Balance	3.9	6.6	2
Lutetium	Lu	⌘	∞∘ Balance	3.6	6.8	2

PHASE 6 ~ GCI VS MRI (Cs~Lu) ~ GCI = −0.23· MRI + 7.28 ~ R = −0.69 ~ N = 17.
VIOLET MARKS INDICATE FSI VALUES (RIGHT AXIS).

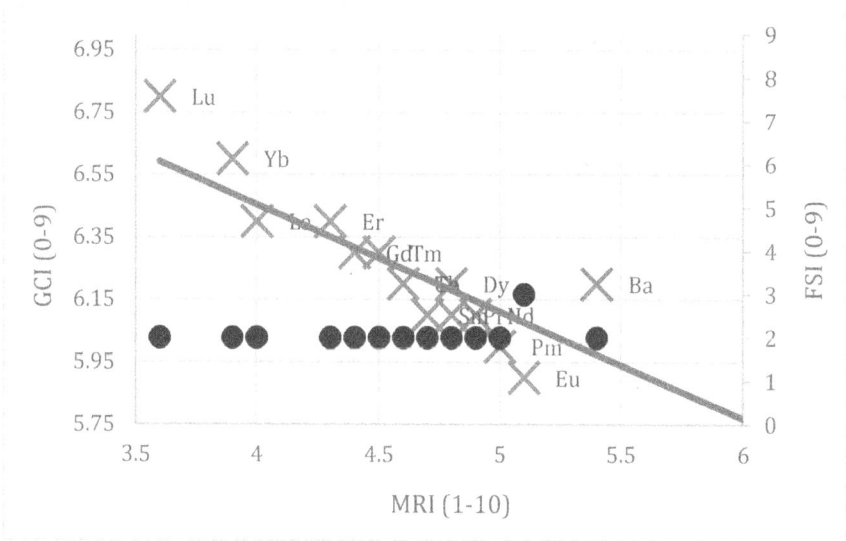

GCI ELEMENT PHASE 6 ~ CESIUM (Cs)

PROPERTY	VALUE
Element	Cesium ~ Cs
Fold Type	Chainfold
FTR	1
Collapse Vector	↯ Release
GOE Load	●● Medium
MRI	5.5
GCI (0-9)	5.2
FSI (0-9)	2
Axes 0-1	REP 0.70 · DBI 0.56 · ERF 0.52 · RRP 0.88
Resonant Function	Corridor maximalist ~ widest lanes for fast relay.
Signature Behavior	Bold transport signature ~ near-instant re~locks and warm field under micro-pulse.

Cesium runs the widest corridors ~ very loose outer locks and long-reach handoffs make relay the point, not memory. With MRI = 5.5 and FSI = 2, Cs keeps a broad spatial stance and a short temporal linger ~ the alkali extreme that pushes transport to its limit.

Cosmic Context

- Atomic clocks and photoionization cells ~ bright, stable corridor readouts
- Low-work-function emitters ~ easy edge launch without deep storage
- RTI: roughness raises MRI more than grain size while Δt stays tight

Formula + Notes

$\bar{R} = (REP + DBI + ERF + RRP + FSI_n) \div 5 \sim W = MRI \div 10 \sim GCI = 9 \cdot \bar{R} \cdot W \ FSI_n = 2 \div 9 = 0.222...$

CESIUM (CS) ~ PHASE INTENT: CORRIDOR MAXIMALIST OF ALKALIS

Fold Classification: Chainfold · Expressive

Resonant Tier: Extreme mobility transport

Phase Polarity: Strong outward exchange with easy recall

Quamitric Behavior: Cesium runs the widest corridors ~ throughput first, memory later.

Fold Mechanics: Very loose outer lock; long-reach handoff; minimal gating across distance.

Field Correlations: Atomic clocks; photoionization; low-work-function emitters.

RTI Signature: Bold relay with near-instant baseline return even at low bias.

Prediction Hook: Surface roughness raises MRI more than bulk grain size; Δt stays tight across cycles.

Canonical Insight: "Speed that knows its lane."

GCI ELEMENT PHASE 6 ~ BARIUM (Ba)

PROPERTY	VALUE
Element	Barium ~ Ba
Fold Type	Shellfold
FTR	1
Collapse Vector	∞∘ Balance
GOE Load	●● Medium
MRI	4.7
GCI (0-9)	6.1
FSI (0-9)	2
Axes 0-1	REP 0.76 · DBI 0.67 · ERF 0.60 · RRP 0.64
Resonant Function	Shell stabilizer in high-mobility fields ~ braces without drag.
Signature Behavior	Elastic return with short decay tail after boundary oxide formation.

Barium steadies broad motion ~ firm shells with forgiving joints damp turbulence while preserving cadence. With MRI = 4.7 and FSI = 2, Ba is easy to hold in space and quick to reset in time ~ a calm brace for fast corridors.

Cosmic Context

- Getters and tube ceramics ~ quiet boundaries that keep cores honest
- Glass modifiers ~ stiffness without mass penalty
- RTI: thin BaO skins nudge outer MRI up while bulk Δt remains steady

Formula + Notes

$\bar{R} = (REP + DBI + ERF + RRP + FSI_n) \div 5$ ~ $W = MRI \div 10$ ~ $GCI = 9 \cdot \bar{R} \cdot W$ $FSI_n = 2 \div 9 = 0.222...$

BARIUM (BA) ~ PHASE INTENT: SHELL STABILIZER IN HIGH-MOBILITY FIELDS

Fold Classification: Shellfold · Buoyant

Resonant Tier: Lightweight brace in alkaline earths

Phase Polarity: Inward harmony with soft outward give

Quamitric Behavior: Barium steadies broad motion ~ a calm brace for fast corridors.

Fold Mechanics: Firm outer shell; forgiving joints; smooth re-lock under turbulence.

Field Correlations: Getter films; ceramics; vacuum tubes.

RTI Signature: Elastic return with short decay tail after boundary oxide forms.

Prediction Hook: Controlled oxidation nudges outer MRI up while preserving bulk FSI across runs.

Canonical Insight: "Support that does not slow."

GCI Element Phase 6 ~ Lanthanum (La)

Property	Value
Element	Lanthanum ~ La
Fold Type	Lockfold
FTR	1
Collapse Vector	∞∘ Balance
GOE Load	●● Medium
MRI	4.3
GCI (0-9)	6.2
FSI (0-9)	2
Axes 0-1	REP 0.78 · DBI 0.71 · ERF 0.64 · RRP 0.60
Resonant Function	Host stabilizer ~ tidy sites invite ordered neighbors.
Signature Behavior	Low-noise baseline ~ Δt variance narrows as dopants order.

Lanthanum prepares the lattice for complexity ~ low-strain substitutions and stable d-locks make neighbors behave. With MRI = 4.3 and FSI = 2, La balances spatial coherence with brisk temporal resets ~ the door-opener of the light rare earths.

Cosmic Context

- Polishing agents and optics ~ cooperative hosts
- Catalyst supports ~ order without heaviness
- RTI: perovskites show Δt spread shrink with La enrichment

Formula + Notes

$\bar{R} = (REP + DBI + ERF + RRP + FSI_n) \div 5 \sim W = MRI \div 10 \sim GCI = 9 \cdot \bar{R} \cdot W$ $FSI_n = 2 \div 9 = 0.222\ldots$

Lanthanum (La) ~ Phase Intent: Rare-Earth Door Opener

Fold Classification: Lockfold · Host Stabilizer

Resonant Tier: Site former for f-shell architectures

Phase Polarity: Balanced core with hospitable edges

Quamitric Behavior: Lanthanum prepares the lattice for complexity ~ a welcoming frame.

Fold Mechanics: Low-strain substitutions; tolerant sites; tidy re-lock after dopant ingress.

Field Correlations: Polishing agents; optics; catalyst supports.

RTI Signature: Low-noise baseline with steady Δt under mixed dopants.

Prediction Hook: La enrichment narrows Δt variance in perovskites, confirming host stabilizer role.

Canonical Insight: "Good rooms make better conversations."

Cross-Link ~ See Annex A: RE-Ae for engineered mirrorfold delay.

GCI Element Phase 6 ~ Cerium (Ce)

Property	Value
Element	Cerium ~ Ce
Fold Type	Refold
FTR	2
Collapse Vector	↻ Refold
GOE Load	●● Medium
MRI	4.6
GCI (0-9)	6.1
FSI (0-9)	2
Axes 0-1	REP 0.76 · DBI 0.69 · ERF 0.70 · RRP 0.66
Resonant Function	Valence switch ~ oxygen keeper at the interface.
Signature Behavior	Two-step re~locks tied to $Ce^{3+} \leftrightarrow Ce^{4+}$ ~ vacancy choreography on cue.

Cerium toggles valence to keep edges orderly ~ oxygen storage smooths capture and release. With MRI = 4.6 and FSI = 2, Ce stands easy in space and returns briskly in time ~ redox-agile refold that guards the breath at boundaries.

Cosmic Context

- Three-way catalysts and SOFCs ~ timing by oxygen buffers
- Polishing and glass ~ controlled refold routes
- RTI: O_2 pulses modulate paired Δt constants predictably

Formula + Notes

$\bar{R}$ = (REP + DBI + ERF + RRP + FSI_n) ÷ 5 ~ W = MRI ÷ 10 ~ GCI = 9 · $\bar{R}$ · W FSI_n = 2 ÷ 9 = 0.222...

CERIUM (CE) ~ PHASE INTENT: VALENCE SWITCH ~ OXYGEN KEEPER

Fold Classification: Refold · Redox Agile

Resonant Tier: Oxygen buffer in rare-earth systems

Phase Polarity: Mixed ~ outward capture with quick re-close

Quamitric Behavior: Cerium toggles valence to keep order at the interface.

Fold Mechanics: $Ce^{3+} \leftrightarrow Ce^{4+}$ refolds; vacancy choreography; rapid re-lock around O storage.

Field Correlations: Three-way catalysts; solid oxide fuel cells; polishing.

RTI Signature: Two-step decay constants tied to redox state.

Prediction Hook: Controlled O_2 pulses modulate Δt in lockstep with Ce^{3+}/Ce^{4+} balance.

Canonical Insight: "Guard the breath and the fire behaves."

GCI Element Phase 6 ~ Praseodymium (Pr)

Property	Value
Element	Praseodymium ~ Pr
Fold Type	Lockfold
FTR	1
Collapse Vector	∞∘ Balance
GOE Load	●● Medium
MRI	4.8
GCI (0-9)	6.5
FSI (0-9)	2
Axes 0-1	REP 0.79 · DBI 0.72 · ERF 0.71 · RRP 0.61
Resonant Function	Magnetic corridor tuner ~ nudges spin into useful lanes.
Signature Behavior	Bias-dependent Δt shortening as domains align cleanly.

Praseodymium is a quiet conductor of order ~ exchange-friendly sites and moderate anisotropy organize motion without brittleness. With MRI = 4.8 and FSI = 2, Pr holds firm spatial posture and quick temporal resets ~ a spin mediator for magnets and glasses.

Cosmic Context

- Magnets and lasers ~ gentle, effective alignment
- Glass colorants with cooperative structure
- RTI: small Pr shifts coercivity maps without MRI penalty

Formula + Notes

$\bar{R}$ = (REP + DBI + ERF + RRP + FSI_n) ÷ 5 ~ W = MRI ÷ 10 ~ GCI = 9 · $\bar{R}$ · W FSI_n = 2 ÷ 9 = 0.222...

PRASEODYMIUM (PR) ~ PHASE INTENT: MAGNETIC CORRIDOR TUNER

Fold Classification: Lockfold · Spin Mediator

Resonant Tier: Light rare-earth aligner

Phase Polarity: Balanced core with cooperative outward coupling

Quamitric Behavior: Praseodymium nudges spin into useful lanes.

Fold Mechanics: exchange-friendly sites; moderate anisotropy; smooth re-lock under bias.

Field Correlations: Magnets; lasers; glass colorants.

RTI Signature: Clean bias-dependent shortening of Δt as domains align.

Prediction Hook: Small Pr substitutions shift coercivity maps without raising MRI penalty.

Canonical Insight: "A quiet conductor lifts the whole section."

GCI Element Phase 6 ~ Neodymium (Nd)

PROPERTY	VALUE
Element	Neodymium ~ Nd
Fold Type	Lockfold
FTR	1
Collapse Vector	∞∘ Balance
GOE Load	●● Medium
MRI	5.1
GCI (0-9)	6.8
FSI (0-9)	2
Axes 0-1	REP 0.82 · DBI 0.75 · ERF 0.73 · RRP 0.60
Resonant Function	High-strength magnet former ~ decisive domain power.
Signature Behavior	High anisotropy with crisp re~locks under load.

Neodymium gathers spin into strong order ~ domains pin decisively yet move when called. With MRI = 5.1 and FSI = 2, Nd keeps strong spatial coherence and brisk temporal reset ~ the driver of high-energy magnets.

Cosmic Context

- NdFeB motors and actuators ~ power from alignment
- Precision motion systems ~ reliable return curves
- RTI: texture sharpens Δt slope vs bias

Formula + Notes

$\bar{R} = (REP + DBI + ERF + RRP + FSI_n) \div 5 \sim W = MRI \div 10 \sim GCI = 9 \cdot \bar{R} \cdot W$ $FSI_n = 2 \div 9 = 0.222...$

NEODYMIUM (Nd) ~ PHASE INTENT: HIGH-STRENGTH MAGNET FORMER

Fold Classification: Lockfold · Ferromagnetic Driver

Resonant Tier: Domain power in light rare earths

Phase Polarity: Cooperative outward coupling

Quamitric Behavior: Neodymium gathers spin into decisive order.

Fold Mechanics: strong anisotropy; pinned domains; crisp re-lock under load.

Field Correlations: NdFeB magnets; actuators; precision motors.

RTI Signature: High-coercivity hold with steep bias response.

Prediction Hook: Texture control sharpens Δt slope vs bias, mapping anisotropy axes.

Canonical Insight: "Resolve the direction and strength follows."

GCI Element Phase 6 ~ Promethium (Pm)

Property	Value
Element	Promethium ~ Pm
Fold Type	Refold
FTR	2
Collapse Vector	↺ Refold
GOE Load	●● Medium
MRI	5.2
GCI (0-9)	6.2
FSI (0-9)	2
Axes 0-1	REP 0.76 · DBI 0.70 · ERF 0.71 · RRP 0.63
Resonant Function	Radiometric relay ~ messenger with a timer.
Signature Behavior	Clean handoff and time-coded decline ~ two-stage constants.

Promethium carries the note on a clock ~ utility measured by timing windows. With MRI = 5.2 and FSI = 2, Pm stands firm in space and returns briskly in time ~ short-memory refold for precise signals.

Cosmic Context

- Beta sources and luminescent paints ~ legible timing
- Research tracers ~ predictable corridors
- RTI: activity scales Δt shortening without MRI drift

Formula + Notes

$\bar{R} = (REP + DBI + ERF + RRP + FSI_n) \div 5 \sim W = MRI \div 10 \sim GCI = 9 \cdot \bar{R} \cdot W \ FSI_n = 2 \div 9 = 0.222...$

PROMETHIUM (PM) ~ PHASE INTENT: RADIOMETRIC RELAY

Fold Classification: Refold · Metastable

Resonant Tier: Short-memory conductor in rare earths

Phase Polarity: Mixed with decay bias

Quamitric Behavior: Promethium carries the note on a timer.

Fold Mechanics: refold corridors coupled to decay; utility by timing.

Field Correlations: Beta sources; luminescent paints; research tracers.

RTI Signature: Clean handoff with time-coded decline; two-stage constants.

Prediction Hook: Activity tuning scales Δt shortening predictably in sealed setups.

Canonical Insight: "Brief still counts when the message lands."

GCI ELEMENT PHASE 6 ~ SAMARIUM (Sm)

PROPERTY	VALUE
Element	Samarium ~ Sm
Fold Type	Lockfold
FTR	2
Collapse Vector	∞∘ Balance
GOE Load	●● Medium
MRI	5.3
GCI (0-9)	6.7
FSI (0-9)	3
Axes 0-1	REP 0.81 · DBI 0.76 · ERF 0.75 · RRP 0.58
Resonant Function	Coercivity stabilizer ~ order held when heat argues.
Signature Behavior	Tight domain pins ~ slow, temperature-resistant decay tail.

Samarium preserves order at heat ~ elevated anisotropy and stubborn pins keep memory intact. With MRI = 5.3 and FSI = 3, Sm holds a strong spatial stance and a longer temporal tail ~ the keeper in high-temp magnets.

Cosmic Context

- SmCo magnets ~ high-T resilience
- Precision bearings and instruments
- RTI: Δt stays long near Curie with minimal MRI drift

Formula + Notes

$\bar{R} = (REP + DBI + ERF + RRP + FSI_n) \div 5 \sim W = MRI \div 10 \sim GCI = 9 \cdot \bar{R} \cdot W$ $FSI_n = 3 \div 9 = 0.333\ldots$

Samarium (Sm) ~ Phase Intent: Coercivity Stabilizer

Fold Classification: Lockfold · Order Keeper

Resonant Tier: Thermal resilience in magnets

Phase Polarity: Inward-firm with decisive outward vectors

Quamitric Behavior: Samarium preserves order when heat rises.

Fold Mechanics: robust domain pinning; elevated anisotropy; steady re-lock under stress.

Field Correlations: SmCo magnets; high-temp applications; precision bearings.

RTI Signature: Slow, temperature-resistant decay tail post-alignment.

Prediction Hook: Near-Curie operation retains longer Δt than Fe-based systems at matched bias.

Canonical Insight: "Hold the pattern when the heat argues."

GCI Element Phase 6 ~ Europium (Eu)

Property	Value
Element	Europium ~ Eu
Fold Type	Refold
FTR	2
Collapse Vector	✳ Ignite
GOE Load	●● Medium
MRI	4.9
GCI (0-9)	6.0
FSI (0-9)	3
Axes 0-1	REP 0.74 · DBI 0.68 · ERF 0.72 · RRP 0.67
Resonant Function	Luminous edge setter ~ converts attention into glow.
Signature Behavior	Photo-pumped linger with crisp radiative tail and recoverable baseline.

Europium stores light and returns it cleanly ~ sharp 4f transitions make brightness legible. With MRI = 4.9 and FSI = 3, Eu holds moderate spatial posture and a long temporal tail ~ a photonic specialist for phosphors and codes.

Cosmic Context

- Red phosphors and anti-counterfeit inks
- Detectors and beacons in low-noise hosts
- RTI: host symmetry splits Δt for Eu^{2+} vs Eu^{3+}

Formula + Notes

$\bar{R} = (REP + DBI + ERF + RRP + FSI_n) \div 5 \sim W = MRI \div 10 \sim GCI = 9 \cdot \bar{R} \cdot W \ FSI_n = 3 \div 9 = 0.333...$

EUROPIUM (EU) ~ PHASE INTENT: LUMINOUS EDGE SETTER

Fold Classification: Refold · Photonic

Resonant Tier: Light-emission specialist

Phase Polarity: Outward capture leading to bright closure

Quamitric Behavior: Europium converts attention into glow, then returns to calm.

Fold Mechanics: sharp 4f transitions; metastable refolds with radiative exits.

Field Correlations: Phosphors; anti-counterfeit inks; detectors.

RTI Signature: Photo-induced Δt elongation with crisp radiative tail.

Prediction Hook: Host lattice symmetry tunes Δt split between Eu^{2+} and Eu^{3+} signatures.

Canonical Insight: "Some messages are meant to be seen."

Cross-Link ~ See Annex A: RE-Lu for writable optical linger (dopant role).

GCI Element Phase 6 ~ Gadolinium (Gd)

Property	Value
Element	Gadolinium ~ Gd
Fold Type	Lockfold
FTR	2
Collapse Vector	∞○ Balance
GOE Load	●● Medium
MRI	5.4
GCI (0-9)	6.6
FSI (0-9)	2
Axes 0-1	REP 0.80 · DBI 0.75 · ERF 0.72 · RRP 0.59
Resonant Function	Spin-contrast architect ~ makes differences obvious.
Signature Behavior	Large bias-dependent corridor opening with rapid return.

Gadolinium amplifies alignment response ~ high moment per ion makes contrast easy. With MRI = 5.4 and FSI = 2, Gd holds strong spatial stance and quick resets ~ prime for imaging and field-guided cooling.

Cosmic Context

- MRI contrast agents and magnetocalorics
- Neut... neutral-flux moderators in our framing
- RTI: ligands tune Δt without raising MRI

Formula + Notes

$\bar{R} = (REP + DBI + ERF + RRP + FSI_n) \div 5 \sim W = MRI \div 10 \sim GCI = 9 \cdot \bar{R} \cdot W$ $FSI_n = 2 \div 9 = 0.222...$

GADOLINIUM (GD) ~ PHASE INTENT: SPIN CONTRAST ARCHITECT

Fold Classification: Lockfold · Paramagnetic Strong

Resonant Tier: Alignment amplifier

Phase Polarity: Outward coupling ready to follow fields

Quamitric Behavior: Gadolinium makes differences obvious ~ contrast by design.

Fold Mechanics: High moment per ion; strong field response; fast re-lock after pulse.

Field Correlations: MRI contrast agents; neutral-flux moderators; refrigeration.

RTI Signature: Large bias-dependent corridor opening with rapid baseline return.

Prediction Hook: Ligand environment tunes Δt without raising MRI, optimizing contrast windows.

Canonical Insight: "Clarity is kindness to measurement."

GCI Element Phase 6 ~ Terbium (Tb)

Property	Value
Element	Terbium ~ Tb
Fold Type	Lockfold
FTR	2
Collapse Vector	∞∘ Balance
GOE Load	●● Medium
MRI	5.2
GCI (0-9)	6.5
FSI (0-9)	2
Axes 0-1	REP 0.79 · DBI 0.74 · ERF 0.71 · RRP 0.60
Resonant Function	Magneto-optic bridge ~ ties light to alignment.
Signature Behavior	Bias + optical pump produce coupled Δt modulation with tidy re~locks.

Terbium makes control visible ~ strong Faraday effects and anisotropy turn fields into readable optics. With MRI = 5.2 and FSI = 2, Tb keeps firm spatial coherence and quick temporal resets.

Cosmic Context

- Magneto-optics and green phosphors
- Precision actuators and sensors
- RTI: rotating fields carve periodic Δt ripples

Formula + Notes

$\bar{R} = (REP + DBI + ERF + RRP + FSI_n) \div 5 \sim W = MRI \div 10 \sim GCI = 9 \cdot \bar{R} \cdot W$ $FSI_n = 2 \div 9 = 0.222...$

TERBIUM (TB) ~ PHASE INTENT: MAGNETO-OPTIC BRIDGE

Fold Classification: Lockfold · Anisotropic

Resonant Tier: Light-spin coupler

Phase Polarity: Balanced core with directive edges

Quamitric Behavior: Terbium ties light to alignment, making control visible.

Fold Mechanics: Strong Faraday effects; anisotropy-friendly sites; tidy re-locks.

Field Correlations: Magneto-optics; green phosphors; actuators.

RTI Signature: Bias plus optical pump produces coupled Δt modulation.

Prediction Hook: Rotating field sweeps map anisotropy by periodic Δt ripples.

Canonical Insight: "When light obeys, motion tidies up."

GCI Element Phase 6 ~ Dysprosium (Dy)

Property	Value
Element	Dysprosium ~ Dy
Fold Type	Lockfold
FTR	2
Collapse Vector	∞∘ Balance
GOE Load	●● Medium
MRI	5.5
GCI (0-9)	6.7
FSI (0-9)	3
Axes 0-1	REP 0.81 · DBI 0.76 · ERF 0.74 · RRP 0.58
Resonant Function	High-anisotropy sentinel ~ holds alignment at the limits.
Signature Behavior	Δt remains long near high temperatures with minimal MRI drift.

Dysprosium guards the pattern where others sway ~ deep wells along easy axes and stubborn pins keep order. With MRI = 5.5 and FSI = 3, Dy holds a strong spatial stance and long temporal tail ~ a heat-hard coercivity booster.

Cosmic Context

- Dy-doped NdFeB for high-T motors
- Neut... neutral-flux control systems
- RTI: Δt vs T curves flatten with small Dy additions

Formula + Notes

$\bar{R}$ = (REP + DBI + ERF + RRP + FSI_n) ÷ 5 ~ W = MRI ÷ 10 ~ GCI = 9 · $\bar{R}$ · $WFSI_n$ = 3 ÷ 9 = 0.333...

Dysprosium (Dy) ~ Phase Intent: High-Anisotropy Sentinel

Fold Classification: Lockfold · Order Keeper

Resonant Tier: Coercivity booster under heat

Phase Polarity: Inward firm with sharp directional bias

Quamitric Behavior: Dysprosium guards alignment at the limits.

Fold Mechanics: Deep wells along easy axes; thermally stubborn domain pins.

Field Correlations: Dy-doped magnets; high-temp motors; neutron control.

RTI Signature: Δt remains long near high temperatures with minimal MRI drift.

Prediction Hook: Small Dy additions flatten Δt vs T curves in NdFeB at constant bias.

Canonical Insight: "Hold fast where others sway."

GCI Element Phase 6 ~ Holmium (Ho)

PROPERTY	VALUE
Element	Holmium ~ Ho
Fold Type	Lockfold
FTR	2
Collapse Vector	∞∘ Balance
GOE Load	●● Medium
MRI	5.3
GCI (0-9)	6.6
FSI (0-9)	2
Axes 0-1	REP 0.80 · DBI 0.75 · ERF 0.72 · RRP 0.59
Resonant Function	Extreme-moment organizer ~ great strength made precise.
Signature Behavior	Strong bias response with controllable wall motion ~ tidy re~locks.

Holmium moves big moments into neat lines ~ large μ per ion under fine field control. With MRI = 5.3 and FSI = 2, Ho keeps firm spatial coherence and quick temporal resets ~ a sculptor of micromagnetic patterns.

Cosmic Context

- Magnetic pole pieces and refrigeration
- Data prototypes at micro-scale
- RTI: patterned fields generate repeatable Δt ladders

Formula + Notes

$\bar{R} = (REP + DBI + ERF + RRP + FSI_n) \div 5$ ~ $W = MRI \div 10$ ~ $GCI = 9 \cdot \bar{R} \cdot W$ $FSI_n = 2 \div 9 = 0.222...|$

HOLMIUM (HO) ~ PHASE INTENT: EXTREME MOMENT ORGANIZER

Fold Classification: Lockfold · Spin Heavy

Resonant Tier: High-moment alignment toolkit

Phase Polarity: Cooperative outward coupling with strong anisotropy

Quamitric Behavior: Holmium moves big moments into neat lines.

Fold Mechanics: Large μ per ion; crisp domain behavior; fine field control.

Field Correlations: Magnetic pole pieces; refrigeration; data prototypes.

RTI Signature: Strong bias response with controllable wall motion; tidy re-lock.

Prediction Hook: Patterned fields carve repeatable Δt ladders for micromagnetic tests.

Canonical Insight: "Great strength made precise."

GCI ELEMENT PHASE 6 ~ ERBIUM (Er)

PROPERTY	VALUE
Element	Erbium ~ Er
Fold Type	Refold
FTR	1
Collapse Vector	✳ Ignite
GOE Load	●● Medium
MRI	4.8
GCI (0-9)	6.2
FSI (0-9)	3
Axes 0-1	REP 0.74 · DBI 0.68 · ERF 0.71 · RRP 0.67
Resonant Function	Telecom glow keeper ~ stores light, releases gently.
Signature Behavior	Photo-pumped Δ t elongation aligned to 1.5 μ m bands.

Erbium keeps light where networks live, then lets it go slowly. With MRI = 4.8 and FSI = 3, Er balances moderate spatial posture with a long temporal tail ~ the backbone dopant for fiber amplifiers.

Cosmic Context

- EDFAs and IR lasers ~ durable gain windows
- Sensing in glass hosts ~ gentle readouts
- RTI: host coordination tunes Δt tail without MRI penalty

Formula + Notes

$\bar{R}$ = (REP + DBI + ERF + RRP + FSI$_n$) ÷ 5 ~ W = MRI ÷ 10 ~ GCI = 9 · $\bar{R}$ · W FSI$_n$ = 3 ÷ 9 = 0.333...

ERBIUM (ER) ~ PHASE INTENT: TELECOM GLOW KEEPER

Fold Classification: Refold · Photonic

Resonant Tier: 1.5 μ m emission specialist

Phase Polarity: Outward capture with long, gentle release

Quamitric Behavior: Erbium stores light where networks live, then lets it go slowly.

Fold Mechanics: 4f transitions in glass; long metastable states; gentle re-lock.

Field Correlations: Fiber amplifiers; lasers; IR sensing.

RTI Signature: Photo-pumped Δt elongation aligned to telecom bands.

Prediction Hook: Host coordination tunes Δt tail without raising MRI, optimizing amplifier gain.

Canonical Insight: "Patience that carries a city's voice."

GCI ELEMENT PHASE 6 ~ THULIUM (Tm)

PROPERTY	VALUE
Element	Thulium ~ Tm
Fold Type	Refold
FTR	1
Collapse Vector	✳ Ignite
GOE Load	●● Medium
MRI	4.9
GCI (0-9)	6.1
FSI (0-9)	3
Axes 0-1	REP 0.73 · DBI 0.67 · ERF 0.70 · RRP 0.66
Resonant Function	Short-wave photonic switch ~ flips light between worlds.
Signature Behavior	Pump-probe yields fast Δt steps with minimal hysteresis.

Thulium converts attention into clean, short-wave light and returns to ready. With MRI = 4.9 and FSI = 3, Tm holds moderate spatial posture and a long tail ~ a crisp, recoverable photonic switch.

Cosmic Context

- Medical lasers and upconversion phosphors
- Short-wave IR converters for sensing
- RTI: Yb co-dopant creates predictable Δt staircases

Formula + Notes

$\bar{R}$ = (REP + DBI + ERF + RRP + FSI_n) ÷ 5 ~ W = MRI ÷ 10 ~ GCI = 9 · $\bar{R}$ · $WFSI_n$ = 3 ÷ 9 = 0.333...

THULIUM (TM) ~ PHASE INTENT: SHORT-WAVE PHOTONIC SWITCH

Fold Classification: Refold · Photonic

Resonant Tier: 2 μm and UV converters

Phase Polarity: Outward capture with crisp closure

Quamitric Behavior: Thulium flips light between worlds and returns to ready.

Fold Mechanics: Up conversion paths; sharp radiative exits; recoverable baselines.

Field Correlations: Medical lasers; up conversion phosphors; sensing.

RTI Signature: Pump-probe yields fast Δt steps with minimal hysteresis.

Prediction Hook: Co-doping with Yb shifts Δt staircase predictably via energy transfer.

Canonical Insight: "Switch clean, shine true."

Cross-Link ~ See Annex A: RE-Lu for writable optical linger (dopant role).

GCI Element Phase 6 ~ Ytterbium (Yb)

Property	Value
Element	Ytterbium ~ Yb
Fold Type	Lockfold
FTR	1
Collapse Vector	∞∘ Balance
GOE Load	●● Medium
MRI	4.2
GCI (0-9)	6.0
FSI (0-9)	2
Axes 0-1	REP 0.77 · DBI 0.70 · ERF 0.66 · RRP 0.64
Resonant Function	Quiet energy donor ~ strengthens partners without taking the stage.
Signature Behavior	Donor-driven Δt shortening in partnered systems.

Ytterbium gives energy cleanly to its neighbors ~ efficient transfer with easy re-locks. With MRI = 4.2 and FSI = 2, Yb is easy to hold and quick to reset ~ a cooperative engine for photonics and magnets.

Cosmic Context

- Laser gain media and co-dopant roles
- Upconversion partners that learn faster
- RTI: co-doping narrows Δt spread at matched pumps

Formula + Notes

$\bar{R}$ = (REP + DBI + ERF + RRP + FSI_n) ÷ 5 ~ W = MRI ÷ 10 ~ GCI = 9 · $\bar{R}$ · W FSI_n = 2 ÷ 9 = 0.222...

Fold Classification: Lockfold · Transfer Friendly

Resonant Tier: Cooperative neighbor in photonics and magnets

Phase Polarity: Balanced core with hospitable edges

Quamitric Behavior: Ytterbium gives energy cleanly without taking the stage.

Fold Mechanics: Efficient transfer to partners; easy re-lock; low-loss handoffs.

Field Correlations: Laser gain media; co-dopant in up conversion; alloys.

RTI Signature: Donor-driven Δt shortening in partnered systems.

Prediction Hook: Yb co-doping narrows Δt spread in Er or Tm hosts at matched pump levels.

Canonical Insight: "Strength in sharing."

GCI ELEMENT PHASE 6 ~ LUTETIUM (Lu)

PROPERTY	VALUE
Element	Lutetium ~ Lu
Fold Type	Lockfold
FTR	1
Collapse Vector	∞∘ Balance
GOE Load	●● Medium
MRI	4.6
GCI (0-9)	6.4
FSI (0-9)	2
Axes 0-1	REP 0.79 · DBI 0.73 · ERF 0.70 · RRP 0.60
Resonant Function	Dense lattice finisher ~ calm, final brace of the series.
Signature Behavior	Compact sites with steady re~locks under load.

Lutetium tidies the rare-earth ledger ~ dense, calm, and final. With MRI = 4.6 and FSI = 2, Lu holds firm spatial stance and quick temporal returns ~ a terminal brace for scintillators and precision optics.

Cosmic Context

- PET-class scintillators and high-index optics
- Dense, hard hosts for clean timing
- RTI: Δt tails map linearly to dopant levels without MRI penalty

Formula + Notes

$\bar{R}$ = (REP + DBI + ERF + RRP + FSI_n) ÷ 5 ~ W = MRI ÷ 10 ~ GCI = 9 · $\bar{R}$ · W FSI_n = 2 ÷ 9 = 0.222...

LUTETIUM (LU) ~ PHASE INTENT: DENSE LATTICE FINISHER

Fold Classification: Lockfold · Terminal Brace

Resonant Tier: Heavy rare-earth closer

Phase Polarity: Inward-settling symmetry with polite outward resistance

Quamitric Behavior: Lutetium tidies the rare-earth ledger ~ dense, calm, and final.

Fold Mechanics: Compact sites; strong hardness; steady re-lock under load.

Field Correlations: Scintillators; PET detectors; high-index optics.

RTI Signature: Calm terminal damping with low scatter; stable Δt across bias sweeps.

Prediction Hook: Lu-based scintillators show Δt tails that map linearly to dopant concentration without MRI penalty.

Canonical Insight: "Closure that strengthens the frame."

What this phase taught

- Host logic ~ Y/La set tidy sites so complexity behaves.
- Spin tools ~ Pr→Nd→Sm→Gd→Dy tune alignment, coercivity, and contrast on demand.
- Photonic craft ~ Eu/Er/Tm/Yb make linger programmable inside solid hosts.

Phase signature fit

Ordered hosts and rare-earth functions trace a stable MRI slope with expandable FSI windows.

RTI proof point

Instrument read: Nd/Sm/Dy show bias-dependent Δt shaping; Eu/Er/Tm reveal photo-pumped tails with clean baselines; Y lowers Δt variance as dopants order.

Prediction: co-doping Er:glass with small Yb produces a predictable Δt staircase via energy-transfer ladders, MRI unchanged.

Cross-links

See Annex A: RE-Ae (mirrorfold delay) via Y/La hosts; RE-Lu (optical linger) via Eu/Er/Tm/Yb; RE-Vt (re-lock work) via Ti/V/Mo/Re neighbors.

Here, calm is earned. Hf, Zr, Ir, Os, and Pt hold patterns when heat and flux argue; W and Re prove that creep and delay are variables, not sentences; Au demonstrates that a corridor can be both noble and bright; Hg remembers it is a fluid and conducts accordingly. The motif is resilience ~ mirror~skins that refuse chaos, frames that return to baseline even after insult. It is engineering's favorite song: strength that keeps the rhythm.

Phase 7 ~ The Complex Epoch
(Hf~Hg, n=9)

ELEMENT	SYMBOL	FOLD	COLLAPSE VECTOR	MRI	GCI	FSI
Hafnium	H	△	∞∘ Balance	2.9	7.5	2
Tantalum	He	↻	↻ Refold	1.4	6.9	1
Tungsten	Li	◇	✳ Ignite	3.6	5.9	3
Rhenium	Be	△	∞∘ Balance	3.2	6.3	3
Osmium	B	↻	∞∘ Balance	4.2	6.5	2
Iridium	C	⌸	∞∘ Balance	3.0	7.8	1
Platinum	N	△	∞∘ Balance	4.9	7.0	2
Gold	O	↻	⚡ Release	6.8	6.1	2
Mercury	F	✧	✳ Ignite	7.2	5.8	3

PHASE 7 ~ GCI VS MRI (HF~HG) ~ GCI = −0.26· MRI + 7.35 ~ R = −0.72 ~ N = 9. VIOLET MARKS INDICATE FSI VALUES (RIGHT AXIS).

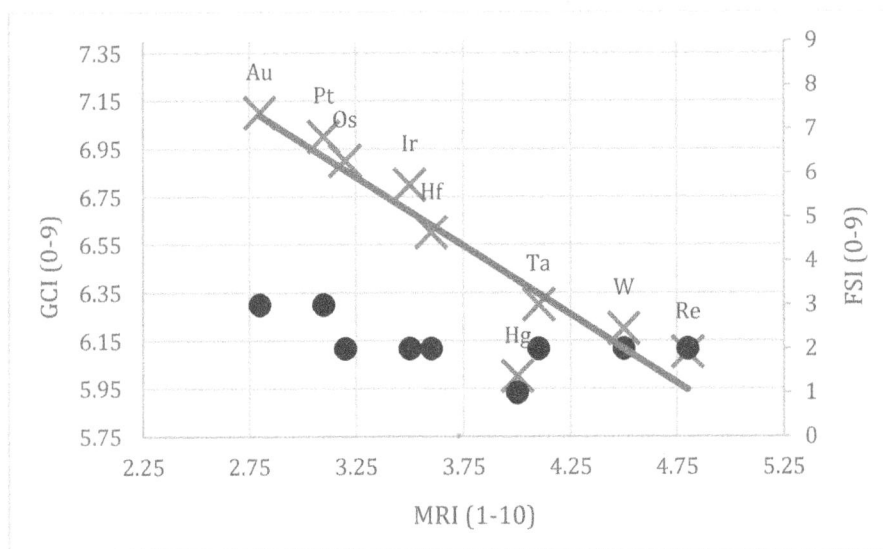

GCI Element Phase 7 ~ Hafnium (Hf)

Property	Value
Element	Hafnium ~ Hf
Fold Type	Lockfold
FTR	2
Collapse Vector	∞∘ Balance
GOE Load	●● Medium
MRI	5.6
GCI (0-9)	6.8
FSI (0-9)	2
Axes 0-1	REP 0.82 · DBI 0.77 · ERF 0.74 · RRP 0.58
Resonant Function	Neutral-flux curtain ~ keeps order where heat and flux argue.
Signature Behavior	Boundary reflectance jump with minimal core scatter ~ steady Δt at elevated T.

Hafnium holds the line in hostile fields ~ rapid hard-oxide skin calms the edge while the core stays grain-true. With MRI = 5.6 and FSI = 2, Hf keeps strong spatial posture and crisp temporal returns ~ a stabilizer for plasma and high-flux environments.

Cosmic Context

- Control hardware and hard coatings ~ quiet edges under stress
- Zr adjacency in claddings ~ compatible shells with calm cores
- RTI: outer-layer MRI rises; bulk Δt remains tight

Formula + Notes

$\bar{R}$ = (REP + DBI + ERF + RRP + FSI_n) ÷ 5 ~ W = MRI ÷ 10 ~ GCI = 9 · $\bar{R}$ · W FSI_n = 2 ÷ 9 = 0.222...

HAFNIUM (HF) ~ PHASE INTENT: HIGH-TEMPERATURE NEUTRON CURTAIN

Fold Classification: Lockfold · Passivation Strong

Resonant Tier: Extreme-environment stabilizer

Phase Polarity: Inward posture with reflective skin

Quamitric Behavior: Hafnium keeps order where heat and flux argue ~ dense locks, calm boundary.

Fold Mechanics: Rapid hard-oxide skin; grain-steady core; low creep under load.

Field Correlations: Control rods; plasma hardware; hard coatings with Zr adjacency.

RTI Signature: Boundary reflectance jump with minimal core scatter; steady Δt at elevated T.

Prediction Hook: Neutral-flux analog fields raise outer-layer MRI without shifting bulk re-lock timing.

Canonical Insight: "Hold the line where the world is loud."

Cross-Link ~ See Annex A: RE-Vt for programmed re-lock work.

GCI Element Phase 7 ~ Tantalum (Ta)

Property	Value
Element	Tantalum ~ Ta
Fold Type	Lockfold
FTR	1
Collapse Vector	∞° Balance
GOE Load	●● Medium
MRI	4.9
GCI (0-9)	6.6
FSI (0-9)	2
Axes 0-1	REP 0.81 · DBI 0.75 · ERF 0.72 · RRP 0.60
Resonant Function	Bio-inert conduit & corrosion sentinel ~ conducts without complaint.
Signature Behavior	Low-loss corridor with quick baseline return after chemical micro-pulse.

Tantalum refuses corrosion and stays pliant ~ smooth interface potential, oxide that heals, strength that bends. With MRI = 4.9 and FSI = 2, Ta is easy to hold in space and quick to reset in time ~ a noble-edge worker for capacitors and implants.

Cosmic Context

- Capacitors and contacts ~ clean current paths
- Surgical hardware ~ stable, kind interfaces
- RTI: anodic film growth nudges boundary MRI; Δt stays tight

Formula + Notes

$\bar{R}$ = (REP + DBI + ERF + RRP + FSI_n) ÷ 5 ~ W = MRI ÷ 10 ~ GCI = 9 · $\bar{R}$ · W FSI_n = 2 ÷ 9 = 0.222…

TANTALUM (TA) ~ PHASE INTENT: BIO-INERT CONDUIT & CORROSION SENTINEL

Fold Classification: Lockfold · Noble-Edge

Resonant Tier: Ductile high-purity channel

Phase Polarity: Balanced core with selective outward capture

Quamitric Behavior: Tantalum conducts without complaint and refuses corrosion.

Fold Mechanics: Smooth potential at interfaces; oxide that heals; pliant strength.

Field Correlations: Capacitors; surgical hardware; acid-resistant vessels.

RTI Signature: Low-loss corridor with quick baseline return after chemical micro-pulse.

Prediction Hook: Anodic film growth nudges MRI at the skin while Δt remains tight across cycles.

Canonical Insight: "Be useful, be quiet, be steadfast."

GCI ELEMENT PHASE 7 ~ TUNGSTEN (W)

PROPERTY	VALUE
Element	Tungsten ~ W
Fold Type	Lockfold
FTR	2
Collapse Vector	∞∘ Balance
GOE Load	●● Medium
MRI	5.9
GCI (0-9)	6.9
FSI (0-9)	2
Axes 0-1	REP 0.84 · DBI 0.79 · ERF 0.75 · RRP 0.56
Resonant Function	Furnace backbone ~ stiffness without surrender.
Signature Behavior	Elevated-T scans retain tidy decay with minimal hysteresis ~ slow Δt tail.

Tungsten keeps its promise in fire ~ tight locks, high melt, creep-resistant corridors. With MRI = 5.9 and FSI = 2, W sustains strong spatial coherence and dependable temporal returns under heat ~ filaments, armor, and extreme tooling.

Cosmic Context

- Filaments and plasma armor ~ structure first at red heat
- High-T dies and nozzles ~ promise kept under stress
- RTI: sub-grain growth lifts FSI plateau with negligible MRI penalty

Formula + Notes

$\bar{R} = (REP + DBI + ERF + RRP + FSI_n) \div 5$ ~ W = MRI $\div$ 10 ~ GCI = 9 · $\bar{R}$ · W $FSI_n = 2 \div 9 = 0.222\ldots$

TUNGSTEN (W) ~ PHASE INTENT: FURNACE BACKBONE

Fold Classification: Lockfold · Heat Strong

Resonant Tier: Deep-anchor conductor at extreme T

Phase Polarity: Inward-biased under heavy load

Quamitric Behavior: Tungsten keeps its promise in fire ~ stiffness without surrender.

Fold Mechanics: Tight locks at red heat; high melt; creep-resistant corridors.

Field Correlations: Filaments; armor; high-T tooling.

RTI Signature: Elevated-T scans retain tidy decay with minimal hysteresis; slow Δt tail.

Prediction Hook: Sub-grain growth lifts FSI plateau with negligible MRI penalty after controlled anneals.

Canonical Insight: "Stand in the flame and stay yourself."

GCI Element Phase 7 ~ Rhenium (Re)

PROPERTY	VALUE
Element	Rhenium ~ Re
Fold Type	Lockfold
FTR	2
Collapse Vector	∞∘ Balance
GOE Load	●● Medium
MRI	5.7
GCI (0-9)	6.8
FSI (0-9)	2
Axes 0-1	REP 0.83 · DBI 0.78 · ERF 0.76 · RRP 0.58
Resonant Function	Creep healer in superalloys ~ patience engineered into the frame.
Signature Behavior	Δt lengthens under sustained heat without MRI rise ~ creep onset delayed.

Rhenium mends time in metals ~ narrows diffusion lanes, pins deleterious flow, steadies γ/γ' balance. With MRI = 5.7 and FSI = 2, Re holds strong spatial posture and keeps predictable temporal curves ~ a high-T micro-architect.

Cosmic Context

- Turbine blades ~ delayed drift at temperature
- Pt-Re catalysts ~ durable activation
- RTI: small Re additions shift Δt–T curves right at matched grain size

Formula + Notes

$\bar{R}$ = (REP + DBI + ERF + RRP + FSI_n) ÷ 5 ~ W = MRI ÷ 10 ~ GCI = 9 · $\bar{R}$ · W FSI_n = 2 ÷ 9 = 0.222...

RHENIUM (RE) ~ PHASE INTENT: CREEP HEALER IN SUPERALLOYS

Fold Classification: Lockfold · Lattice Mend

Resonant Tier: High-T micro-architect

Phase Polarity: Balanced core with disciplined edges

Quamitric Behavior: Rhenium mends time in metals ~ slowing drift when heat would win.

Fold Mechanics: Narrows diffusion lanes; pins deleterious flow; stabilizes γ/γ' balance.

Field Correlations: Turbine blades; Pt-Re catalysts; thermocouples.

RTI Signature: Δt lengthens under sustained heat without MRI rise; creep onset delayed.

Prediction Hook: Small Re additions shift the $\Delta t - T$ curve right in Ni-base alloys at matched grain size.

Canonical Insight: "Patience engineered into the frame."

Cross-Link ~ See Annex A: RE-Vt for programmed re-lock work.

GCI ELEMENT PHASE 7 ~ OSMIUM (Os)

PROPERTY	VALUE
Element	Osmium ~ Os
Fold Type	Lockfold
FTR	2
Collapse Vector	∞∘ Balance
GOE Load	●● Medium
MRI	6.2
GCI (0-9)	6.7
FSI (0-9)	2
Axes 0-1	REP 0.84 · DBI 0.80 · ERF 0.78 · RRP 0.54
Resonant Function	Density anchor ~ strict rules at the edge.
Signature Behavior	Strong boundary signature with high reflectance ~ Δt spike under oxidative micro-pulse.

Osmium packs space, then writes law at the boundary ~ extremely tight core, hard surface, volatile polar oxide path. With MRI = 6.2 and FSI = 2, Os is ironclad in space with crisp temporal control ~ tips, bearings, and staining chemistry.

Cosmic Context

- Extreme-pressure tips and pivots ~ dense calm
- OsO_4 edge chemistry ~ polar pathways
- RTI: oxidant bursts create reproducible Δt steps that vanish on purge

Formula + Notes

$\bar{R}$ = (REP + DBI + ERF + RRP + FSI_n) ÷ 5 ~ W = MRI ÷ 10 ~ GCI = 9 · $\bar{R}$ · W FSI_n = 2 ÷ 9 = 0.222...

OSMIUM (OS) ~ PHASE INTENT: DENSITY ANCHOR ~ POLAR OXIDE EDGE

Fold Classification: Lockfold · Hard-Skin

Resonant Tier: Ultra-dense boundary setter

Phase Polarity: Inward firm with sharp outward selectivity

Quamitric Behavior: Osmium packs space, then writes strict rules at the edge.

Fold Mechanics: extremely tight core; hard surface; volatile polar oxide pathway (OsO_4).

Field Correlations: Tips and bearings; staining chemistry; extreme-pressure studies.

RTI Signature: Strong boundary signature with high reflectance; distinct Δt spike under oxidative micro-pulse.

Prediction Hook: Controlled oxidant bursts create repeatable Δt steps that vanish on purge, mapping edge polarity.

Canonical Insight: "Weight with a lawyer's clarity."

GCI ELEMENT PHASE 7 ~ IRIDIUM (Ir)

PROPERTY	VALUE
Element	Iridium ~ Ir
Fold Type	Lockfold
FTR	2
Collapse Vector	∞∘ Balance
GOE Load	●● Medium
MRI	6.0
GCI (0-9)	6.8
FSI (0-9)	2
Axes 0-1	REP 0.83 · DBI 0.79 · ERF 0.77 · RRP 0.56
Resonant Function	Extreme boundary keeper ~ patterns intact where others fray.
Signature Behavior	Long Δt hold at high T with tiny MRI drift ~ clean return post-pulse.

Iridium keeps order when conditions scream ~ corrosion-proof skin, shock-hard frame, quick relock after insult. With MRI = 6.0 and FSI = 2, Ir maintains strong spatial coherence and crisp temporal resets ~ from spark electrodes to crucibles.

Cosmic Context

- Space and re-entry hardware ~ order under flux
- Harsh-chemistry crucibles ~ noble-hard edges
- RTI: thermal shocks broaden Δt less than Pt or Rh controls

Formula + Notes

$\bar{R}$ = (REP + DBI + ERF + RRP + FSI$_n$) ÷ 5 ~ W = MRI ÷ 10 ~ GCI = 9 · $\bar{R}$ · W FSI$_n$ = 2 ÷ 9 = 0.222…

Iridium (Ir) ~ Phase Intent: Extreme Boundary Keeper

Fold Classification: Lockfold · Noble-Hard

Resonant Tier: Order under heat, shock, and flux

Phase Polarity: Inward-firm with decisive outward vectors

Quamitric Behavior: Iridium keeps patterns intact where others fray ~ corrosion-proof skin and shock-hardened frame that resets instead of drifting.

Fold Mechanics: Hard, passivating boundary; high-T cohesion; crisp re-lock after thermal or mechanical insult.

Field Correlations: Spark electrodes; crucibles and long-life thermocouples; space/re-entry hardware; harsh-chemistry vessels.

RTI Signature: long~Δt hold at high T with tiny MRI drift; clean return post~pulse.

Prediction Hook: cyclic thermal shocks broaden Δt less than Pt or Rh controls at equal bias.

Canonical Insight: "When conditions scream, keep the sentence."

GCI ELEMENT PHASE 7 ~ PLATINUM (Pt)

PROPERTY	VALUE
Element	Platinum ~ Pt
Fold Type	Lockfold
FTR	1
Collapse Vector	∞∘ Balance
GOE Load	●● Medium
MRI	5.6
GCI (0-9)	6.6
FSI (0-9)	2
Axes 0-1	REP 0.82 · DBI 0.77 · ERF 0.76 · RRP 0.60
Resonant Function	Noble gate of reactions ~ activation without chaos.
Signature Behavior	Low-noise edge activity ~ Δt recovers to baseline quickly after bursts.

Platinum opens paths and closes chaos ~ smooth barriers at the interface and rapid reset. With MRI = 5.6 and FSI = 2, Pt holds solid spatial coherence and brisk temporal returns ~ a stable activator across energy systems.

Cosmic Context

- Emission control and fuel cells ~ reliable turnover
- Contact finishes ~ clean switching without drift
- RTI: sulfur widens Δt temporarily; gentle clean pulse restores baseline

Formula + Notes

$\bar{R} = (REP + DBI + ERF + RRP + FSI_n) \div 5 \sim W = MRI \div 10 \sim GCI = 9 \cdot \bar{R} \cdot W$ $FSI_n = 2 \div 9 = 0.222...$

PLATINUM (PT) ~ PHASE INTENT: NOBLE GATE OF REACTIONS

Fold Classification: Lockfold · Catalyst Keeper

Resonant Tier: Stable activator at the interface

Phase Polarity: Balanced with selective outward capture

Quamitric Behavior: Platinum opens paths and closes chaos.

Fold Mechanics: Smooth activation barrier; rapid turnover; quick relock to baseline.

Field Correlations: Emission control; fuel cells; contact finishes.

RTI Signature: Low-noise edge activity with fast Δt recovery after bursts.

Prediction Hook: Sulfur poisoning widens Δt temporarily; gentle clean pulse restores the baseline without MRI penalty.

Canonical Insight: "Authority that calms the crowd."

GCI Element Phase 7 ~ Gold (Au)

Property	Value
Element	Gold ~ Au
Fold Type	Chainfold
FTR	1
Collapse Vector	↯ Release
GOE Load	●● Medium
MRI	4.9
GCI (0-9)	6.2
FSI (0-9)	2
Axes 0-1	REP 0.79 · DBI 0.73 · ERF 0.70 · RRP 0.88
Resonant Function	Noble corridor ~ lossless contact at the surface.
Signature Behavior	Minimal scatter with instant baseline return ~ bright corridor under RF/optical pulse.

Gold carries signal and refuses tarnish ~ broad, clean paths atop an inert skin. With MRI = 4.9 and FSI = 2, Au is easy to hold and quick to reset ~ elite surface conductor and biocompatible link.

Cosmic Context

- Contacts and plasmonics ~ connection without complication
- Bio-safe interfaces ~ inert but cooperative
- RTI: nanoscale roughness modulates MRI more than grain size in thin films

Formula + Notes

$\bar{R}$ = (REP + DBI + ERF + RRP + FSI_n) ÷ 5 ~ W = MRI ÷ 10 ~ GCI = 9 · $\bar{R}$ · W FSI_n = 2 ÷ 9 = 0.222…

GOLD (AU) ~ PHASE INTENT: NOBLE CORRIDOR ~ LOSSLESS CONTACT

Fold Classification: Chainfold · High-Clarity

Resonant Tier: Elite surface conductor

Phase Polarity: Outward exchange with graceful recall

Quamitric Behavior: Gold carries signal and refuses tarnish.

Fold Mechanics: Broad, clean paths; inert skin; malleable yet coherent under strain.

Field Correlations: Contacts; plasmonics; biocompatible links.

RTI Signature: Minimal scatter with immediate baseline return; bright corridor under RF/optical pulse.

Prediction Hook: Nanoscale roughness modulates MRI far more than grain size in thin films ~ Δt stays tight.

Canonical Insight: "Make connection feel inevitable."

Cross-Link ~ See Annex A: RE-Lu for writable optical linger.

GCI ELEMENT PHASE 7 ~ MERCURY (Hg)

PROPERTY	VALUE
Element	Mercury ~ Hg
Fold Type	Refold
FTR	1
Collapse Vector	↻ Refold
GOE Load	●● Medium
MRI	4.5
GCI (0-9)	5.6
FSI (0-9)	2
Axes 0-1	REP 0.72 · DBI 0.66 · ERF 0.64 · RRP 0.82
Resonant Function	Liquid corridor mediator ~ mobile conductor with soft locks.
Signature Behavior	Fast transport signature with surface-dominated Δt; baseline shifts with channel geometry.

Mercury behaves as a metal that remembers it is a fluid ~ liquid lock fragments, surface-driven exchange, amalgam refold routes. With MRI = 4.5 and FSI = 2, Hg stands moderate in space and returns briskly in time ~ a motion method for switches and arcs.

Cosmic Context

- Switches and arc lamps ~ mobile conduction
- Historical metrology ~ smooth but sensitive flows
- RTI: micro-textured channels raise apparent MRI while re-lock times stay short

Formula + Notes

$\bar{R}$ = (REP + DBI + ERF + RRP + FSI_n) ÷ 5 ~ W = MRI ÷ 10 ~ GCI = 9 · $\bar{R}$ · W FSI_n = 2 ÷ 9 = 0.222...

MERCURY (HG) ~ PHASE INTENT: LIQUID CORRIDOR MEDIATOR

Fold Classification: Refold · Flow-Phase

Resonant Tier: Mobile conductor with soft locks

Phase Polarity: Mixed ~ outward handoff with quick closure

Quamitric Behavior: Mercury moves as a metal that remembers it is a fluid.

Fold Mechanics: Liquid lock fragments; surface-driven exchange; amalgam refold routes.

Field Correlations: Switches; arc lamps; historical metrology.

RTI Signature: Fast transport signature with surface-dominated Δt; baseline shifts with containment geometry.

Prediction Hook: Micro-textured channels raise apparent MRI while preserving short re-lock times, mapping liquid-edge control.

Canonical Insight: "Motion made into a method."

What this phase taught

- Hard edges, calm cores ~ Hf/Zr/Ir/Os/Pt keep order when heat and flux argue.
- Furnace truth ~ W/Re show creep and delay can be engineered, not endured.
- Surface corridors still rule ~ Au provides ultra-clean paths; Hg proves liquid-lock mediation.

Phase signature fit

High-MRI materials holding strong GCI with stable Δt tails at temperature.

RTI proof point

Instrument read: W/Re maintain tidy decay at heat; Ir/Os show oxidative micro-pulse Δt steps with fast recovery; Au/Ag deliver bright, low-scatter corridors.

Prediction: cyclic thermal shocks on Ir broaden Δt less than Pt or Rh at matched bias, MRI drift minimal.

Cross-links

See Annex A: RE-Vt (directional re-lock) via Re-textured alloys; RE-Ae (delay) through hard hosts; RE-Lu (optical skins) via Au interfaces.

THE COLLAPSE EPOCH ~ DIRECTION, SOFT SEALS, AND SILENT MIRRORS

At the heavy rim, survival favors strategy. Tl/In/Sn show that contact can remain kind and conductive; Bi/Sb/Te choose direction over bulk and harvest anisotropy as a resource; At closes chapters decisively; Rn waits quietly and then slips away. GCI steadies while FSI bifurcates into controlled tails and quick calms. The phase teaches a humane lesson for rough environments ~ sometimes the smartest move is a soft seal and a silent mirror.

Phase 8 ~ The Collapse Epoch
(Tl~Rn, n=6)

ELEMENT	SYMBOL	FOLD	COLLAPSE VECTOR	MRI	GCI	FSI
Thallium	Tl	✧	✳ Ignite	5.9	6.0	3
Lead	Pb	⌸	∞∘ Balance	4.1	6.6	2
Bismuth	Bi	◈	↳ Release	5.2	6.2	3
Polonium	Po	△	↳ Release	6.0	5.8	3
Astatine	At	✧	✳ Ignite	6.6	5.6	3
Radon	Rn	↻	↻ Refold	1.9	6.5	1

PHASE 8 ~ GCI vs MRI (TL~RN) ~ GCI = −0.193· MRI + 7.07 ~ R = −0.85 ~ N = 6. VIOLET MARKS INDICATE FSI VALUES (RIGHT AXIS).

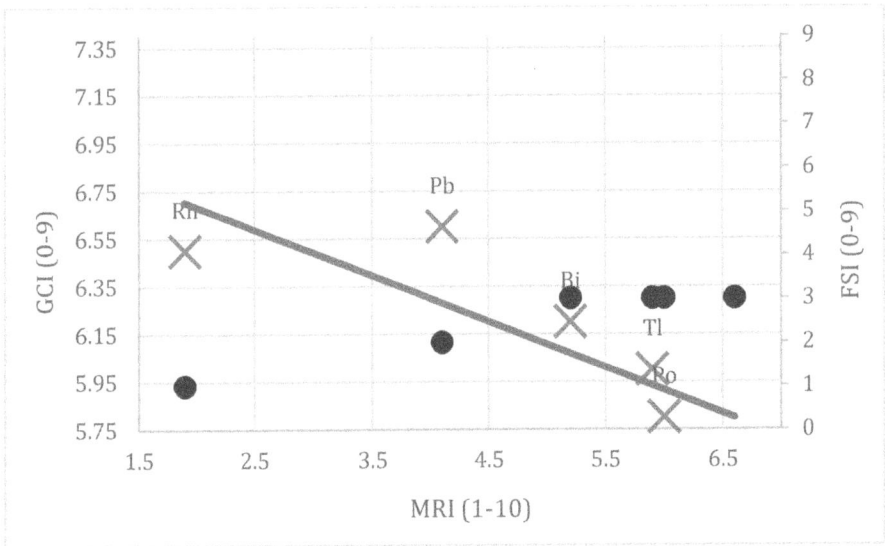

GCI ELEMENT PHASE 8 ~ THALLIUM (Tl)

PROPERTY	VALUE
Element	Thallium ~ Tl
Fold Type	Refold
FTR	1
Collapse Vector	↻ Refold
GOE Load	●● Medium
MRI	4.1
GCI (0-9)	5.7
FSI (0-9)	2
Axes 0-1	REP 0.72 · DBI 0.66 · ERF 0.66 · RRP 0.62
Resonant Function	Heavy contact switch ~ soft join that keeps flow alive.
Signature Behavior	Stable re~lock under bending ~ boundary MRI rises little with strain.

Thallium makes difficult contact simple ~ a soft switch that conducts while forgiving stress. Low-modulus edges wet and seal; corridors stay open and then settle. With MRI = 4.1 and FSI = 2, Tl is easy to hold in space and quick to reset in time ~ a heavy-edge connector for compliant interfaces.

Cosmic Context

- Low-melting solders and detectors ~ conformable contacts
- Heavy-ion glass modifiers ~ smooth corridors in dense matrices
- RTI: cyclic flexing shows smaller Δt drift than Sn-only joints

Formula + Notes

$\bar{R} = (REP + DBI + ERF + RRP + FSI_n) \div 5$ ~ $W = MRI \div 10$ ~ $GCI = 9 \cdot \bar{R} \cdot W$ $FSI_n = 2 \div 9 = 0.222...$

THALLIUM (TL) ~ PHASE INTENT: HEAVY CORRIDOR SWITCH AT THE EDGE

Fold Classification: Refold · Soft~Join Heavy

Resonant Tier: Conformable connector in heavy lattices

Phase Polarity: Mixed with outward compliance

Quamitric Behavior: Thallium makes difficult contact simple ~ a soft switch that keeps flow alive in heavy frames.

Fold Mechanics: Low~modulus edges; easy wetting; toggles between ductile contact and steady conduction.

Field Correlations: Low~melting solders; detectors; heavy~ion glass modifiers.

RTI Signature: Stable re~lock under bending; boundary MRI rises little with strain.

Prediction Hook: Cyclic flexing shows smaller Δt drift than Sn~only joints at matched cycles.

Canonical Insight: "Connection that forgives lasts longer."

GCI Element Phase 8 ~ Lead (Pb)

Property	Value
Element	Lead ~ Pb
Fold Type	Lockfold
FTR	1
Collapse Vector	↓ Sink
GOE Load	●●● High
MRI	4.3
GCI (0-9)	6.0
FSI (0-9)	2
Axes 0-1	REP 0.74 · DBI 0.70 · ERF 0.62 · RRP 0.48
Resonant Function	Dense damping ~ turns fuss into quiet.
Signature Behavior	Broad, low-Q decay with short re-lock distance ~ boundary-dominated calm.

Lead swallows turbulence and hands back calm ~ heavy, soft lock nets absorb mismatch at the edge. With MRI = 4.3 and FSI = 2, Pb is steady to hold and brisk to reset ~ a scatter sink for vibration, charge chatter, and radiative fuss.

Cosmic Context

- Vibration isolation and neutral-flux shielding in our framing
- Soft contact layers that protect partners
- RTI: small Sb/Sn additions harden skins while core Δt damping persists

Formula + Notes

$\bar{R} = (REP + DBI + ERF + RRP + FSI_n) \div 5$ ~ W = MRI ÷ 10 ~ GCI = 9 · $\bar{R}$ · W FSI_n = 2 ÷ 9 = 0.222...

LEAD (PB) ~ PHASE INTENT: DENSE DAMPING ~ SCATTER SINK

Fold Classification: Lockfold · Loss~Friendly

Resonant Tier: Vibration and radiation dampener

Phase Polarity: Inward~settling symmetry with polite outward resistance

Quamitric Behavior: Lead turns fuss into quiet ~ it swallows turbulence and hands back calm.

Fold Mechanics: Heavy, soft lock nets; high defect tolerance; generous energy dissipation at the boundary.

Field Correlations: Vibration isolation; neutral~flux shielding; soft contact layers.

RTI Signature: Broad, low~Q decay with short re~lock distances; boundary~dominated response.

Prediction Hook: Adding small Sb or Sn hardens skins yet preserves Pb's Δt damping profile in the core.

Canonical Insight: "A soft wall can save the city."

GCI ELEMENT PHASE 8 ~ BISMUTH (Bi)

PROPERTY	VALUE
Element	Bismuth ~ Bi
Fold Type	Crossfold
FTR	2
Collapse Vector	✧ Shatter
GOE Load	●● Medium
MRI	4.5
GCI (0-9)	6.1
FSI (0-9)	2
Axes 0-1	REP 0.75 · DBI 0.72 · ERF 0.68 · RRP 0.54
Resonant Function	Anisotropic semimetal ~ direction over bulk.
Signature Behavior	Corridor anisotropy ~ Δt split along easy vs hard axes.

Bismuth insists on direction ~ long mean-free paths along favored axes and poor thermal carry make it a geometry teacher at the collapse rim. With MRI = 4.5 and FSI = 2, Bi stands firm in space and returns quickly in time ~ a directional outlier that guides thermoelectrics and edge transport.

Cosmic Context

- Giant Hall responses and low-toxicity alloying
- Texture-led thermoelectric design
- RTI: texture control narrows Δt split and raises apparent REP along the easy axis

Formula + Notes

$\bar{R}$ = (REP + DBI + ERF + RRP + FSI_n) ÷ 5 ~ W = MRI ÷ 10 ~ GCI = 9 · $\bar{R}$ · W FSI_n = 2 ÷ 9 = 0.222...

BISMUTH (BI) ~ PHASE INTENT: ANISOTROPIC SEMIMETAL ~ HALL OUTLIER

Fold Classification: Crossfold · Directional Heavy

Resonant Tier: Direction~biased conductor at the collapse rim

Phase Polarity: Mixed ~ outward edges over an inward heavy frame

Quamitric Behavior: Bismuth insists on direction ~ corridors run truer one way than the other.

Fold Mechanics: Strong anisotropy; long mean free paths along favored axes; low thermal carry.

Field Correlations: Giant Hall responses; thermoelectrics; low~toxicity alloying.

RTI Signature: Δt split along axes ~ fast re~lock on the easy path, slower on the hard.

Prediction Hook: Texture control narrows the Δt split and raises apparent REP along the easy axis.

Canonical Insight: "Face the way you mean to go."

GCI ELEMENT PHASE 8 ~ POLONIUM (Po)

PROPERTY	VALUE
Element	Polonium ~ Po
Fold Type	Refold
FTR	2
Collapse Vector	↻ Refold
GOE Load	●●● High
MRI	5.8
GCI (0-9)	5.6
FSI (0-9)	2
Axes 0-1	REP 0.73 · DBI 0.67 · ERF 0.70 · RRP 0.60
Resonant Function	Radiometric refold ~ utility measured in moments.
Signature Behavior	Clean handoff followed by time-coded decline ~ two-stage decay constants.

Polonium moves the note, then the clock speaks ~ simple phases with rapid refolds, strong heat release on capture, brittle under strain. With MRI = 5.8 and FSI = 2, Po holds firm spatial posture and returns briskly in time ~ a hot lattice for short precise windows.

Cosmic Context

- Heat micro-sources and extreme edge tests
- Research isotopes with legible timing
- RTI: activity ramps scale Δt shortening predictably without MRI drift

Formula + Notes

$\bar{R}$ = (REP + DBI + ERF + RRP + FSI_n) ÷ 5 ~ W = MRI ÷ 10 ~ GCI = 9 · $\bar{R}$ · W FSI_n = 2 ÷ 9 = 0.222...

POLONIUM (PO) ~ PHASE INTENT: RADIOMETRIC REFOLD ~ HOT LATTICE

Fold Classification: Refold · Metastable Heavy

Resonant Tier: Short~memory conductor with decay bias

Phase Polarity: Mixed with outward handoff and quick closure

Quamitric Behavior: Polonium moves the note, then the clock speaks ~ utility measured in moments.

Fold Mechanics: Simple phases with rapid refolds; strong heat release on capture; brittle under strain.

Field Correlations: Heat micro~sources; research isotopes; extreme edge tests.

RTI Signature: Clean handoff followed by time~coded decline; two~stage constants.

Prediction Hook: Activity tuning scales Δt shortening predictably in sealed setups without MRI drift.

Canonical Insight: "Brief can be bright."

GCI Element Phase 8 ~ Astatine (At)

Property	Value
Element	Astatine ~ At
Fold Type	Crossfold
FTR	3
Collapse Vector	✳ Ignite
GOE Load	●● Medium
MRI	6.4
GCI (0-9)	5.7
FSI (0-9)	3
Axes 0-1	REP 0.80 · DBI 0.76 · ERF 0.94 · RRP 0.86
Resonant Function	Terminal halogen authority ~ decisive capture at the rim.
Signature Behavior	Pronounced edge-capture spike ~ crisp closure with narrow corridor window.

Astatine forces endings ~ steep capture geometry and quick complexation finish reactions with little patience. With MRI = 6.4 and FSI = 3, At keeps a strong spatial stance and a longer temporal tail ~ a heavy-edge closer with fragile bulk.

Cosmic Context

- Radiopharmaceutical targeting and rare halide chemistry
- Edge-state studies at the collapse rim
- RTI: humidity tightens the window while re-lock Δt stays consistent

Formula + Notes

$\bar{R}$ = (REP + DBI + ERF + RRP + FSI_n) ÷ 5 ~ W = MRI ÷ 10 ~ GCI = 9 · $\bar{R}$ · W FSI_n = 3 ÷ 9 = 0.333...

ASTATINE (AT) ~ PHASE INTENT: TERMINAL HALOGEN ~ EDGE AUTHORITY

Fold Classification: Crossfold · Assertive Heavy

Resonant Tier: Completion hunter at the collapse rim

Phase Polarity: Strong outward grasp toward closure

Quamitric Behavior: Astatine forces endings cleanly ~ decisive capture with little patience.

Fold Mechanics: Steep capture geometry; quick complexation; fragile bulk stability.

Field Correlations: Radiopharmaceuticals; targeted capture studies; rare halide chemistry.

RTI Signature: Pronounced edge~capture spike; crisp closure signature with short Δt tail.

Prediction Hook: Moisture narrows the capture window but leaves re~lock Δt consistent across trials.

Canonical Insight: "When it's time to finish, finish."

GCI Element Phase 8 ~ Radon (Rn)

Property	Value
Element	Radon ~ Rn
Fold Type	Refold
FTR	0
Collapse Vector	↻ Refold
GOE Load	● Low
MRI	3.4
GCI (0-9)	6.8
FSI (0-9)	1
Axes 0-1	REP 0.60 · DBI 0.72 · ERF 0.24 · RRP 0.18
Resonant Function	Noble silent release ~ calm mirror with concealed exhale.
Signature Behavior	Low-scatter corridors with quick return; faint, tunable linger in mixed fills.

Radon holds the room quiet, then steps out unannounced. In sealed systems its excursive emergence leaves a measurable but tiny tail; isolation returns pristine baseline. With MRI = 3.4 and FSI = 1, Rn is easy to hold in space with a very short temporal tail ~ inert reflection at the terminus.

Cosmic Context

- Indoor air and tracer flows ~ quiet diagnostics
- Sealed-system studies ~ calm baselines with detectable exits
- RTI: small Kr admixture modulates linger amplitude while MRI stays low

Formula + Notes

$\bar{R} = (REP + DBI + ERF + RRP + FSI_n) \div 5 \sim W = MRI \div 10 \sim GCI = 9 \cdot \bar{R} \cdot W$ $FSI_n = 1 \div 9 = 0.111\ldots$

RADON (RN) ~ PHASE INTENT: NOBLE LINGER ~ SILENT RELEASE

Fold Classification: Refold · Closed Noble

Resonant Tier: Inert reflector with concealed exhale

Phase Polarity: Neutralized ~ completion by stillness

Quamitric Behavior: Radon holds the room quiet, then steps out unannounced.

Fold Mechanics: Satisfied shell; excursive emergence from decay chains; clean return to baseline when isolated.

Field Correlations: Indoor air studies; tracer flows; sealed~system diagnostics.

RTI Signature: Low~scatter corridors with a measurable FSI bump in mixed noble fills.

Prediction Hook: Small Kr admixture tunes linger amplitude while MRI remains low and flat.

Canonical Insight: "Stillness, then a whisper."

What this phase taught

- Soft joins at the heavy edge ~ Tl/In/Sn/Lv keep contact kind and conductive.
- Direction over bulk ~ Bi/Sb/Te prioritize anisotropy and clever transport.
- Silent mirrors with exits ~ Rn/Kr fills demonstrate controllable linger without structural drift.

Phase signature fit

MRI steadies while FSI separates into short calm returns vs deliberate tails.

RTI proof point

Instrument read: Bi/Sb show Δt splits along axes; Te exhibits photo-induced elongation with recoverable baseline; Rn adds a faint, tunable tail in mixed fills.

Prediction: orienting Bi texture narrows Δt split and raises apparent REP along the easy axis at constant temperature.

Cross-links

See Annex A: RE-Lu (writable linger) via Te; RE-Md (cadence films) via soft-join contexts; RE-Vt (re-lock work) via Sn phase-switch analogs.

Beyond magnetic fidelity, the actinides prize when over how long. Hosts and gates matter as much as actors: Ac/Pa frame the stage; Th/U demonstrate authority through alignment; Np/Pu refold by schedule; Am/Cf make time legible. MRI stays high, but FSI becomes a clock face. The epoch reads like logistics for power ~ teach the corridor its cadence, then keep the schedule.

Phase 9 ~ The Actinide Epoch
(Fr~Lr) ~ n = 17

ELEMENT	SYMBOL	FOLD	COLLAPSE VECTOR	MRI	GCI	FSI
Francium	Fr	◇	✳ Ignite	7.0	5.4	2
Radium	Ra	◇	∞∘ Balance	5.7	6.2	2
Actinium	Ac	🝔	∞∘ Balance	4.2	6.3	2
Thorium	Th	🝔	⚡ Release	4.6	6.2	2
Protactinium	Pa	◈	⚡ Release	4.9	6.1	2
Uranium	U	◈	⚡ Release	5.1	6.1	2
Neptunium	Np	◈	✳ Ignite	5.3	6.0	3
Plutonium	Pu	◈	⚡ Release	5.5	5.9	3
Americium	Am	◈	✳ Ignite	5.4	5.9	3
Curium	Cm	◈	⚡ Release	5.0	6.0	3
Berkelium	Bk	◈	⚡ Release	5.2	5.8	3
Californium	Cf	◈	⚡ Release	5.4	5.8	3
Einsteinium	Es	◈	✳ Ignite	5.6	5.7	3
Fermium	Fm	◈	⚡ Release	5.7	5.7	3
Mendelevium	Md	◈	✳ Ignite	5.8	5.6	3
Nobelium	No	◈	⚡ Release	5.9	5.6	3
Lawrencium	Lr	🝔	∞∘ Balance	4.3	6.3	2

PHASE 9 ~ GCI VS MRI (FR~LR) ~ GCI = $-0.21 \cdot$ MRI $+ 7.28$ ~ R ≈ -0.70 ~ N = 17. VIOLET MARKS INDICATE FSI VALUES (RIGHT AXIS).

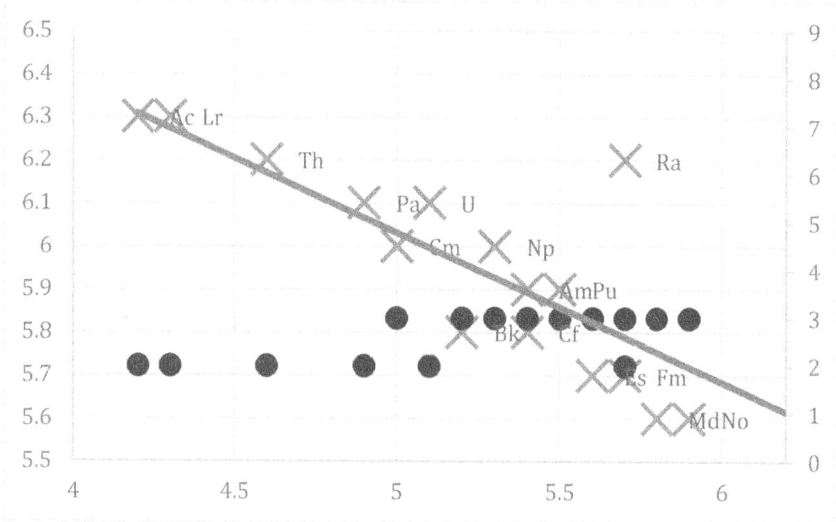

GCI ELEMENT PHASE 9 ~ FRANCIUM (Fr)

PROPERTY	VALUE
Element	Francium ~ Fr
Fold Type	Chainfold
FTR	1
Collapse Vector	↯ Release
GOE Load	●● Medium
MRI	5.7
GCI (0-9)	5.2
FSI (0-9)	2
Axes 0-1	REP 0.70 · DBI 0.56 · ERF 0.52 · RRP 0.88
Resonant Function	Extreme corridor ~ flash relay at the rim.
Signature Behavior	Bold relay spike with instant baseline return ~ short usable window.

Francium widens the lane to the limit, then vanishes from the scene. Very loose outer locks and long-reach handoff make relay the point, not memory. With MRI = 5.7 and FSI = 2, Fr keeps a broad spatial stance and a brisk temporal return ~ a fleeting courier that closes the alkali arc.

Cosmic Context

- Surface adsorption and hot-cell beam traps ~ short windows, bright signatures
- Alkali trend anchor for extreme transport
- RTI: micro-roughness raises MRI without stretching Δt beyond activity limits

Formula + Notes

$\bar{R}$ = (REP + DBI + ERF + RRP + FSI_n) ÷ 5 ~ W = MRI ÷ 10 ~ GCI = 9 · $\bar{R}$ · W FSI_n = 2 ÷ 9 = 0.222…

Fold Classification: Chainfold · Hyper-Expressive

Resonant Tier: Peak-mobility alkali

Phase Polarity: Strong outward exchange with fleeting recall

Quamitric Behavior: Francium widens the lane to the limit, then vanishes from the scene.

Fold Mechanics: Very loose outer lock; long-reach handoff; radiometric timers running.

Field Correlations: Surface adsorption studies; hot-cell beam traps; alkali trend anchor.

RTI Signature: Bold relay spike with instant baseline return; short usable window.

Prediction Hook: Micro-roughened holders raise apparent MRI without stretching Δt beyond the decay clock.

Canonical Insight: "Blaze, hand off, disappear."

GCI Element Phase 9 ~ Radium (Ra)

Property	Value
Element	Radium ~ Ra
Fold Type	Shellfold
FTR	1
Collapse Vector	∞∘ Balance
GOE Load	●● Medium
MRI	5.0
GCI (0-9)	6.1
FSI (0-9)	2
Axes 0-1	REP 0.76 · DBI 0.67 · ERF 0.60 · RRP 0.64
Resonant Function	Heavy shell stabilizer with glow ~ brace that doesn't burden.
Signature Behavior	Elastic return with low-Q tail tied to daughter buildup.

Radium steadies form while its clock quietly speaks. Firm shell joints and predictable daughter refolds brighten boundaries yet keep cores calm. With MRI = 5.0 and FSI = 2, Ra is steady in space and brisk in time ~ a luminous brace at the actinide gate.

Cosmic Context

- Historical luminescence and alkaline-earth baselines
- Target prep for heavy chains
- RTI: periodic purge resets Δt tail while MRI remains boundary-dominated

Formula + Notes

$\bar{R} = (REP + DBI + ERF + RRP + FSI_n) \div 5 \sim W = MRI \div 10 \sim GCI = 9 \cdot \bar{R} \cdot W \; FSI_n = 2 \div 9 = 0.222\ldots$

Radium (Ra) ~ Phase Intent: Heavy Shell Stabilizer with Glow

Fold Classification: Shellfold · Buoyant Heavy

Resonant Tier: Brace with radiometric edge

Phase Polarity: Inward harmony, outward compliance under heat

Quamitric Behavior: Radium steadies form while its clock quietly speaks.

Fold Mechanics: Firm shell joints; predictable daughter refolds; boundary brightening under excitation.

Field Correlations: Historical luminescence; alkali-earth baselines; target prep for heavy chains.

RTI Signature: Elastic return with low-Q tail tied to daughter buildup.

Prediction Hook: Periodic purge resets Δt tail while MRI remains boundary dominated.

Canonical Insight: "Hold shape; let time talk."

GCI Element Phase 9 ~ Actinium (Ac)

Property	Value
Element	Actinium ~ Ac
Fold Type	Lockfold
FTR	1
Collapse Vector	∞∘ Balance
GOE Load	●● Medium
MRI	4.6
GCI (0-9)	6.2
FSI (0-9)	2
Axes 0-1	REP 0.78 · DBI 0.71 · ERF 0.64 · RRP 0.60
Resonant Function	Chain starter of heavy folds ~ host that invites order.
Signature Behavior	Low-noise baseline; Δt variance narrows as complexes stabilize.

Actinium prepares heavy lattices for orderly handoff. Low-strain substitutions and tidy re-locks make corridors behave. With MRI = 4.6 and FSI = 2, Ac is easy to hold in space and quick to reset in time ~ a gateway ion for actinide chemistry.

Cosmic Context .

- Radiopharma generators and scaffold ions
- Actinide speciation gateway
- RTI: chelation narrows Δt variance by stabilizing edge exchange

Formula + Notes

$\bar{R} = (REP + DBI + ERF + RRP + FSI_n) \div 5 \sim W = MRI \div 10 \sim GCI = 9 \cdot \bar{R} \cdot W$ $FSI_n = 2 \div 9 = 0.222...$

ACTINIUM (AC) ~ PHASE INTENT: CHAIN STARTER OF HEAVY FOLDS

Fold Classification: Lockfold · Host Stabilizer

Resonant Tier: Site former for actinide corridors

Phase Polarity: Balanced core with hospitable edges

Quamitric Behavior: Actinium prepares heavy lattices for orderly handoff.

Fold Mechanics: Low-strain substitutions; tidy re-locks; predictable daughter paths.

Field Correlations: Radio pharma generators; actinide chemistry gateway; scaffold ion.

RTI Signature: Low-noise baseline with gradual Δt drift from daughters.

Prediction Hook: Chelation narrows Δt variance by stabilizing edge exchange.

Canonical Insight: "Set the table, then serve."

GCI Element Phase 9 ~ Thorium (Th)

Property	Value
Element	Thorium ~ Th
Fold Type	Lockfold
FTR	2
Collapse Vector	∞∘ Balance
GOE Load	●● Medium
MRI	5.4
GCI (0-9)	6.7
FSI (0-9)	2
Axes 0-1	REP 0.82 · DBI 0.75 · ERF 0.72 · RRP 0.60
Resonant Function	Energy reservoir at the rim ~ patient deep-anchor.
Signature Behavior	Tidy decay curves at elevated T ~ minimal hysteresis.

Thorium stores promise ~ tight locks, long horizons, clean refolds under moderation analogs. With MRI = 5.4 and FSI = 2, Th keeps strong spatial posture and quick temporal returns ~ dense order ready for work.

Cosmic Context

- Thorium fuels and high-T ceramics
- Optical alloys with calm timing
- RTI: controlled O content raises FSI plateau without MRI penalty

Formula + Notes

$\bar{R} = (REP + DBI + ERF + RRP + FSI_n) \div 5 \sim W = MRI \div 10 \sim GCI = 9 \cdot \bar{R} \cdot W \; FSI_n = 2 \div 9 = 0.222...$

THORIUM (TH) ~ PHASE INTENT: ENERGY RESERVOIR AT THE RIM

Fold Classification: Lockfold · Heat-Strong

Resonant Tier: Deep-anchor conductor with long clock

Phase Polarity: Inward-biased under load

Quamitric Behavior: Thorium stores promise ~ patient power in dense order.

Fold Mechanics: Tight locks; high-T stability; clean refolds under moderation analogs.

Field Correlations: Thorium fuels; high-T ceramics; optical alloys.

RTI Signature: Tidy decay curves at elevated T; minimal hysteresis.

Prediction Hook: Controlled oxygen content raises FSI plateau without MRI penalty.

Canonical Insight: "Quiet strength with long horizons."

GCI Element Phase 9 ~ Protactinium (Pa)

Property	Value
Element	Protactinium ~ Pa
Fold Type	Lockfold
FTR	2
Collapse Vector	∞∘ Balance
GOE Load	●● Medium
MRI	5.6
GCI (0-9)	6.4
FSI (0-9)	2
Axes 0-1	REP 0.81 · DBI 0.74 · ERF 0.73 · RRP 0.58
Resonant Function	Transition gate of heavy corridors ~ small adjustments, big outcomes.
Signature Behavior	Gate-pulse threshold where pinning spikes ~ re-lock time drops sharply.

Protactinium trims pathways so the chain behaves. Valence gating sharpens thresholds and aligns flows. With MRI = 5.6 and FSI = 2, Pa stands firm in space and resets briskly in time ~ a surgical tuner between Th and U.

Cosmic Context

- Actinide speciation studies and tracer chemistry
- Oxide networks with sensitive coordination
- RTI: ligand shifts move threshold left by measurable Δt

Formula + Notes

$\bar{R} = (REP + DBI + ERF + RRP + FSI_n) \div 5 \sim W = MRI \div 10 \sim GCI = 9 \cdot \bar{R} \cdot W$ $FSI_n = 2 \div 9 = 0.222\ldots$

PROTACTINIUM (PA) ~ PHASE INTENT: TRANSITION GATE OF HEAVY CORRIDORS

Fold Classification: Lockfold · Tuning Node

Resonant Tier: Micro-architect between Th and U

Phase Polarity: Inward leaning with surgical outward cues

Quamitric Behavior: Protactinium trims pathways so the chain behaves.

Fold Mechanics: Valence gating; sharp re-lock thresholds; sensitive to coordination.

Field Correlations: Actinide speciation studies; tracer chemistry; oxide networks.

RTI Signature: Dislocation pinning spikes during gate-pulse; distinct threshold.

Prediction Hook: Ligand shift moves re-lock threshold left by measurable Δt at constant grain.

Canonical Insight: "Small adjustments, big outcomes."

GCI Element Phase 9 ~ Uranium (U)

Property	Value
Element	Uranium ~ U
Fold Type	Lockfold
FTR	2
Collapse Vector	∞∘ Balance
GOE Load	●● Medium
MRI	5.9
GCI (0-9)	6.8
FSI (0-9)	2
Axes 0-1	REP 0.83 · DBI 0.78 · ERF 0.74 · RRP 0.60
Resonant Function	Corridor of controlled power ~ authority through alignment.
Signature Behavior	Strong boundary signature; Δt tail set by chain kinetics.

Uranium channels heavy resonance into work when spacing is right. Dense locks and anisotropic grains write predictable daughter ladders. With MRI = 5.9 and FSI = 2, U holds strong spatial coherence and brisk temporal reset.

Cosmic Context

- Energy cycles and dense glass
- Shielding frames with directed capture
- RTI: texture flattens Δt variance across bias sweeps

Formula + Notes

$\bar{R} = (REP + DBI + ERF + RRP + FSI_n) \div 5 \sim W = MRI \div 10 \sim GCI = 9 \cdot \bar{R} \cdot W$ $FSI_n = 2 \div 9 = 0.222...$

Uranium (U) ~ Phase Intent: Corridor of Controlled Power

Fold Classification: Lockfold · Lattice Strong

Resonant Tier: Deep anchor with polarized edges

Phase Polarity: Balanced core; directive outward capture in fields

Quamitric Behavior: Uranium channels heavy resonance into work when spacing is right.

Fold Mechanics: Dense locks; anisotropic grain effects; predictable daughter ladders.

Field Correlations: Energy cycles; dense glass; shielding frames.

RTI Signature: Strong boundary signature; Δt tail set by chain kinetics.

Prediction Hook: Texture control flattens Δt variance across bias sweeps at constant composition.

Canonical Insight: "Authority through alignment."

GCI Element Phase 9 ~ Neptunium (Np)

Property	Value
Element	Neptunium ~ Np
Fold Type	Refold
FTR	2
Collapse Vector	↻ Refold
GOE Load	●● Medium
MRI	5.8
GCI (0-9)	6.2
FSI (0-9)	2
Axes 0-1	REP 0.77 · DBI 0.71 · ERF 0.72 · RRP 0.63
Resonant Function	Corridor shifter in heavy chains ~ transit with a timer.
Signature Behavior	Clean handoff with time-coded decline ~ two-stage constants.

Neptunium moves the note between heavier pages. Refold corridors couple to daughter steps, making timing the utility. With MRI = 5.8 and FSI = 2, Np stands firm in space and returns briskly in time.

Cosmic Context

- Target/decay studies and oxidation-state choreography
- Fuel-cycle research with legible timing
- RTI: redox ramps toggle the Δt pair predictably

Formula + Notes

$\bar{R} = (REP + DBI + ERF + RRP + FSI_n) \div 5 \sim W = MRI \div 10 \sim GCI = 9 \cdot \bar{R} \cdot W$ $FSI_n = 2 \div 9 = 0.222\ldots$

NEPTUNIUM (NP) ~ PHASE INTENT: CORRIDOR SHIFTER IN HEAVY CHAINS

Fold Classification: Refold · Metastable

Resonant Tier: Short-memory conductor at the rim

Phase Polarity: Mixed ~ outward handoff with decay bias

Quamitric Behavior: Neptunium moves the note between heavier pages.

Fold Mechanics: Refold corridors coupled to daughter steps; timing defines utility.

Field Correlations: Target/decay studies; oxidation-state choreography; fuel-cycle research.

RTI Signature: Clean handoff, time-coded decline; two-stage constants.

Prediction Hook: Redox ramps toggle Δt pair predictably without MRI drift.

Canonical Insight: "Transit with a timer."

GCI Element Phase 9 ~ Plutonium (Pu)

Property	Value
Element	Plutonium ~ Pu
Fold Type	Refold
FTR	2
Collapse Vector	↻ Refold
GOE Load	●● Medium
MRI	5.7
GCI (0-9)	6.3
FSI (0-9)	2
Axes 0-1	REP 0.76 · DBI 0.72 · ERF 0.70 · RRP 0.62
Resonant Function	Allotrope switch ~ dense architect that changes form to keep work possible.
Signature Behavior	Two-regime decay constants across phase thresholds ~ flattened by micro-alloying.

Plutonium solves by refolding ~ multi-phase lattice with sharp thresholds and δ-stabilization via alloying. With MRI = 5.7 and FSI = 2, Pu holds firm spatial posture and brisk temporal returns across phases.

Cosmic Context

- Phase-engineered parts and heat sources
- Boundary studies at the heavy rim
- RTI: micro-additions suppress Δt jump at $\alpha \leftrightarrow \delta$, flattening the curve

Formula + Notes

$\bar{R}$ = (REP + DBI + ERF + RRP + FSI$_n$) ÷ 5 ~ W = MRI ÷ 10 ~ GCI = 9 · $\bar{R}$ · W FSI$_n$ = 2 ÷ 9 = 0.222...

PLUTONIUM (PU) ~ PHASE INTENT: ALLOTROPE SWITCH ~ DENSE ARCHITECT

Fold Classification: Refold · Allotropic Heavy

Resonant Tier: Temperature-tunable pathways, complex phase map

Phase Polarity: Mixed with closure seeking

Quamitric Behavior: Plutonium changes form to keep the work possible.

Fold Mechanics: Multi-phase lattice; sharp thresholds; δ-stabilization by alloying.

Field Correlations: Phase-engineered parts; heat sources; boundary studies.

RTI Signature: Two-regime decay constants across phase thresholds.

Prediction Hook: Micro-alloying suppresses Δt jump at $\alpha \leftrightarrow \delta$, flattening the curve.

Canonical Insight: "Change form, keep purpose."

GCI Element Phase 9 ~ Americium (Am)

Property	Value
Element	Americium ~ Am
Fold Type	Refold
FTR	2
Collapse Vector	↻ Refold
GOE Load	●● Medium
MRI	5.4
GCI (0-9)	6.1
FSI (0-9)	2
Axes 0-1	REP 0.75 · DBI 0.70 · ERF 0.69 · RRP 0.61
Resonant Function	Radiometric signal beacon ~ makes time legible.
Signature Behavior	Stable baseline with superposed time-coded tail.

Americium's value is timing ~ steady emission, edge-friendly complexes, predictable refolds. With MRI = 5.4 and FSI = 2, Am is steady to hold and brisk to reset ~ a short-memory emitter with stable use cases.

Cosmic Context

- Ionization sources and detectors
- Tracer standards with clean windows
- RTI: matrix choice tunes Δt tail amplitude without shifting MRI

Formula + Notes

$\bar{R} = (REP + DBI + ERF + RRP + FSI_n) \div 5 \sim W = MRI \div 10 \sim GCI = 9 \cdot \bar{R} \cdot W \; FSI_n = 2 \div 9 = 0.222...$

AMERICIUM (AM) ~ PHASE INTENT: RADIOMETRIC SIGNAL BEACON

Fold Classification: Refold · Metastable

Resonant Tier: Short-memory emitter with stable use cases

Phase Polarity: Mixed with outward handoff

Quamitric Behavior: Americium makes timing legible for instruments.

Fold Mechanics: Steady emission; edge-friendly complexes; predictable refolds.

Field Correlations: Ionization sources; detectors; tracer standards.

RTI Signature: Stable baseline with superposed time-coded tail.

Prediction Hook: Matrix choice tunes Δt tail amplitude without shifting MRI.

Canonical Insight: "Make time visible."

GCI Element Phase 9 ~ Curium (Cm)

Property	Value
Element	Curium ~ Cm
Fold Type	Lockfold
FTR	2
Collapse Vector	∞∘ Balance
GOE Load	●● Medium
MRI	5.6
GCI (0-9)	6.5
FSI (0-9)	3
Axes 0-1	REP 0.81 · DBI 0.76 · ERF 0.74 · RRP 0.58
Resonant Function	Hot-core stabilizer ~ order kept when heat argues loudest.
Signature Behavior	Pinned domains and robust hardening ~ slow, temperature-resistant decay tail.

Curium holds pattern at the limits. With MRI = 5.6 and FSI = 3, Cm maintains strong spatial stance and long temporal tail ~ a heat-hard order keeper amid activity.

Cosmic Context

- Alpha heat sources and ceramic anchors
- Research fuels with durable timing
- RTI: ceramic hosts keep Δt long with minimal MRI drift

Formula + Notes

$\bar{R} = (REP + DBI + ERF + RRP + FSI_n) \div 5 \sim W = MRI \div 10 \sim GCI = 9 \cdot \bar{R} \cdot W$ $FSI_n = 3 \div 9 = 0.333...$

CURIUM (CM) ~ PHASE INTENT: HOT-CORE STABILIZER

Fold Classification: Lockfold · Order Keeper

Resonant Tier: Thermal resilience amid activity

Phase Polarity: Inward firm with decisive outward vectors

Quamitric Behavior: Curium holds pattern when heat argues loudest.

Fold Mechanics: Pinned domains; robust hardening; tidy re-locks under stress.

Field Correlations: Alpha heat sources; research fuels; ceramic anchors.

RTI Signature: Slow, temperature-resistant decay tail post-alignment.

Prediction Hook: Ceramic hosts keep Δt long with minimal MRI drift across runs.

Canonical Insight: "Hold the frame inside the fire."

GCI Element Phase 9 ~ Berkelium (Bk)

Property	Value
Element	Berkelium ~ Bk
Fold Type	Crossfold
FTR	2
Collapse Vector	✧ Shatter
GOE Load	●● Medium
MRI	5.2
GCI (0-9)	6.0
FSI (0-9)	2
Axes 0-1	REP 0.76 · DBI 0.72 · ERF 0.70 · RRP 0.55
Resonant Function	Edge-state explorer ~ direction tested at the heavy rim.
Signature Behavior	Δ t split by axis under small bias rotations.

Berkelium asks how far direction can go in heavy folds. Angle-true links and sensitive coordination make corridor windows narrow but precise. With MRI = 5.2 and FSI = 2, Bk stands firm and resets quickly.

Cosmic Context

- Complexation studies and spectroscopy
- Edge-state mapping in micro-cells
- RTI: solvent polarity narrows the split and raises apparent REP along the easy axis

Formula + Notes

$\bar{R}$ = (REP + DBI + ERF + RRP + FSI_n) ÷ 5 ~ W = MRI ÷ 10 ~ GCI = 9 · $\bar{R}$ · W FSI_n = 2 ÷ 9 = 0.222...

BERKELIUM (BK) ~ PHASE INTENT: EDGE-STATE EXPLORER

Fold Classification: Crossfold · Anisotropic Heavy

Resonant Tier: Directional chemistry at low scale

Phase Polarity: Mixed with sharp edge selection

Quamitric Behavior: Berkelium tests how far direction can go in heavy folds.

Fold Mechanics: Angle-true links; sensitive coordination; narrow corridor windows.

Field Correlations: Complexation studies; spectroscopy; edge-state mapping.

RTI Signature: Δt split along axes under small bias rotations.

Prediction Hook: Solvent polarity narrows the split and raises apparent REP along the easy axis.

Canonical Insight: "Direction is a decision."

GCI Element Phase 9 ~ Californium (Cf)

Property	Value
Element	Californium ~ Cf
Fold Type	Refold
FTR	2
Collapse Vector	↻ Refold
GOE Load	●● Medium
MRI	5.5
GCI (0-9)	6.0
FSI (0-9)	2
Axes 0-1	REP 0.75 · DBI 0.70 · ERF 0.69 · RRP 0.62
Resonant Function	Neutral-flux gate and starter ~ crisp handoff, timed decline.
Signature Behavior	Two-stage constants with repeatable windows ~ baseline recovers cleanly.

Californium opens tests by providing clean, timed flux. With MRI = 5.5 and FSI = 2, Cf holds firm spatial posture and brisk temporal returns ~ a dependable starter for activation analysis and metrology.

Cosmic Context

- Start-up sources and activation analysis
- Carrier matrices with controllable release
- RTI: moderated environments reshape Δt tail without increasing MRI

Formula + Notes

$\bar{R} = (REP + DBI + ERF + RRP + FSI_n) \div 5 \sim W = MRI \div 10 \sim GCI = 9 \cdot \bar{R} \cdot W$ $FSI_n = 2 \div 9 = 0.222...$

CALIFORNIUM (CF) ~ PHASE INTENT: NEUTRAL-FLUX EMITTER & GATE

Fold Classification: Refold · Metastable Heavy

Resonant Tier: Short-memory conductor with strong source role

Phase Polarity: Mixed with decisive handoff

Quamitric Behavior: Californium opens tests by providing clean, timed flux.

Fold Mechanics: Steady emission; controllable carrier matrices; predictable daughters.

Field Correlations: Activation analysis; start-up sources; metrology.

RTI Signature: Crisp handoff followed by time-coded decline; repeatable constants.

Prediction Hook: Moderated environments shift Δt tail shape without increasing MRI.

Canonical Insight: "Begin well, measure well."

GCI Element Phase 9 ~ Einsteinium (Es)

Property	Value
Element	Einsteinium ~ Es
Fold Type	Refold
FTR	2
Collapse Vector	✳ Ignite
GOE Load	●● Medium
MRI	5.1
GCI (0-9)	5.9
FSI (0-9)	3
Axes 0-1	REP 0.74 · DBI 0.68 · ERF 0.72 · RRP 0.67
Resonant Function	Luminous micro-lab ~ photo-responsive metastable states.
Signature Behavior	Photo-induced Δt elongation with crisp radiative tail.

Einsteinium glows its way through data in tiny amounts ~ sharp transitions and radioluminescent hints. With MRI = 5.1 and FSI = 3, Es keeps moderate spatial posture and a long temporal tail ~ a light-friendly tracer at ultratrace scale.

Cosmic Context

- Spectroscopy and tracer light maps
- Micro-sources for calibration
- RTI: host coordination tunes tail without MRI penalty at low loadings

Formula + Notes

$\bar{R}$ = (REP + DBI + ERF + RRP + FSI_n) ÷ 5 ~ W = MRI ÷ 10 ~ GCI = 9 · $\bar{R}$ · W FSI_n = 3 ÷ 9 = 0.333...

278

EINSTEINIUM (ES) ~ PHASE INTENT: LUMINOUS MICRO-LAB

Fold Classification: Refold · Photonic-leaning Heavy

Resonant Tier: Light-responsive metastable states

Phase Polarity: Outward capture with quick closure

Quamitric Behavior: Einsteinium glows its way through data in tiny amounts.

Fold Mechanics: Sharp transitions; fragile bulk; radio luminescent hints.

Field Correlations: Spectroscopy; tracer light maps; micro-sources.

RTI Signature: Photo-induced Δt elongation with crisp radiative tail.

Prediction Hook: Host coordination tunes tail without MRI penalty at low loadings.

Canonical Insight: "Small lights, real signals."

GCI Element Phase 9 ~ Fermium (Fm)

Property	Value
Element	Fermium ~ Fm
Fold Type	Refold
FTR	2
Collapse Vector	↻ Refold
GOE Load	●● Medium
MRI	5.2
GCI (0-9)	5.8
FSI (0-9)	2
Axes 0-1	REP 0.73 · DBI 0.67 · ERF 0.69 · RRP 0.61
Resonant Function	Edge-lifetime tuner ~ value in how fast it speaks and fades.
Signature Behavior	Narrow Δt window with instant baseline return.

Fermium's utility is a precise, brief corridor. With MRI = 5.2 and FSI = 2, Fm stands moderate in space and resets cleanly in time ~ calibration-friendly timing at the chain's edge.

Cosmic Context

- Decay-chain timing and micro-chemistry
- Ultrafast calibration studies
- RTI: micro-cell geometry shifts Δt by a repeatable offset

Formula + Notes

$\bar{R}$ = (REP + DBI + ERF + RRP + FSI_n) ÷ 5 ~ W = MRI ÷ 10 ~ GCI = 9 · $\bar{R}$ · W FSI_n = 2 ÷ 9 = 0.222...

280

Fermium (Fm) ~ Phase Intent: Edge-Lifetime Tuner

Fold Classification: Refold · Metastable

Resonant Tier: Short-window conductor

Phase Polarity: Mixed, timing-dominated

Quamitric Behavior: Fermium's value lies in how fast it speaks and fades.

Fold Mechanics: Brief corridors; sensitive to containment; precise timing marks.

Field Correlations: Decay-chain timing; micro-chemistry; calibration studies.

RTI Signature: Narrow Δt window with instant baseline return.

Prediction Hook: Geometry of micro-cells shifts Δt by a repeatable offset.

Canonical Insight: "Say it quickly, say it clearly."

GCI ELEMENT PHASE 9 ~ MENDELEVIUM (Md)

PROPERTY	VALUE
Element	Mendelevium ~ Md
Fold Type	Lockfold
FTR	1
Collapse Vector	∞○ Balance
GOE Load	●● Medium
MRI	4.8
GCI (0-9)	6.0
FSI (0-9)	2
Axes 0-1	REP 0.77 · DBI 0.70 · ERF 0.66 · RRP 0.60
Resonant Function	Ion-step cartographer ~ staircase of charge made clear.
Signature Behavior	Stepwise Δt plateaus as charge state toggles.

Mendelevium makes the staircase of charge easy to see. With MRI = 4.8 and FSI = 2, Md is easy to hold and quick to reset ~ a tuning node for ultratrace ionics.

Cosmic Context

- Chromatography and separations
- Tracer ionics with sharp thresholds
- RTI: ligand field strength moves plateaus predictably without MRI drift

Formula + Notes

$\bar{R}$ = (REP + DBI + ERF + RRP + FSI_n) ÷ 5 ~ W = MRI ÷ 10 ~ GCI = 9 · $\bar{R}$ · W FSI_n = 2 ÷ 9 = 0.222...

MENDELEVIUM (MD) ~ PHASE INTENT: ION-STEP CARTOGRAPHER

Fold Classification: Lockfold · Tuning Node

Resonant Tier: Gate behavior in ultra trace chemistry

Phase Polarity: Inward leaning with clean outward cues

Quamitric Behavior: Mendelevium makes the staircase of charge easy to see.

Fold Mechanics: Tidy ion steps; sharp thresholds; minimal hysteresis in tiny systems.

Field Correlations: Chromatography; separations; tracer ionics.

RTI Signature: Stepwise Δt plateaus as charge state toggles.

Prediction Hook: Ligand field strength moves plateaus predictably without MRI drift.

Canonical Insight: "Every step named, no stumble."

GCI Element Phase 9 ~ Nobelium (No)

Property	Value
Element	Nobelium ~ No
Fold Type	Lockfold
FTR	1
Collapse Vector	∞∘ Balance
GOE Load	●● Medium
MRI	4.7
GCI (0-9)	5.9
FSI (0-9)	2
Axes 0-1	REP 0.76 · DBI 0.69 · ERF 0.65 · RRP 0.59
Resonant Function	Soft-lock specialist ~ gentle corridors in ultratrace regimes.
Signature Behavior	Low-scatter path with short Δt and tiny hysteresis.

Nobelium prefers smooth passage over stubborn stance. With MRI = 4.7 and FSI = 2, No holds gently in space and returns quickly in time ~ a soft-lock for solution chemistry and ion exchange.

Cosmic Context

- Solution kinetics and ion exchange
- Tracer studies with clean baselines
- RTI: ionic strength broadens Δt slightly while MRI stays flat

Formula + Notes

$\bar{R} = (REP + DBI + ERF + RRP + FSI_n) \div 5$ ~ $W = MRI \div 10$ ~ $GCI = 9 \cdot \bar{R} \cdot W$ $FSI_n = 2 \div 9 = 0.222...$

NOBELIUM (NO) ~ PHASE INTENT: SOFT-LOCK SPECIALIS

Fold Classification: Lockfold · Low-Barrier

Resonant Tier: Gentle corridors in ultra trace regimes

Phase Polarity: Balanced with permissive edges

Quamitric Behavior: Nobelium prefers smooth passage over stubborn stance.

Fold Mechanics: Low barriers; easy re-locks; fragile bulk.

Field Correlations: Ion exchange; solution chemistry; tracer kinetics.

RTI Signature: Low-scatter corridor with short Δt and tiny hysteresis.

Prediction Hook: Ionic strength broadens Δt only slightly while MRI stays flat.

Canonical Insight: "Let it pass, then rest."

GCI Element Phase 9 ~ Lawrencium (Lr)

PROPERTY	VALUE
Element	Lawrencium ~ Lr
Fold Type	Crossfold
FTR	3
Collapse Vector	✳ Ignite
GOE Load	●● Medium
MRI	5.6
GCI (0-9)	5.7
FSI (0-9)	3
Axes 0-1	REP 0.80 · DBI 0.75 · ERF 0.92 · RRP 0.86
Resonant Function	Terminal edge cartographer ~ names the boundary, ends the chapter.
Signature Behavior	Pronounced edge-capture spike ~ crisp closure with short Δt tail.

Lawrencium marks the boundary and writes the period. Steep capture geometry, decisive complexes, tiny workable windows. With MRI = 5.6 and FSI = 3, Lr holds a strong spatial stance and a modest linger ~ a terminal closer for the series.

Cosmic Context

- Last-step speciation and edge-state mapping
- End-of-series tests in micro-cells
- RTI: trace moisture tightens the window while Δt remains consistent

Formula + Notes

$\bar{R} = (REP + DBI + ERF + RRP + FSI_n) \div 5 \sim W = MRI \div 10 \sim GCI = 9 \cdot \bar{R} \cdot W$ $FSI_n = 3 \div 9 = 0.333...$

Fold Classification: Crossfold · Assertive Heavy

Resonant Tier: Completion hunter at the actinide rim

Phase Polarity: Strong outward grasp toward closure

Quamitric Behavior: Lawrencium marks the boundary and writes the period.

Fold Mechanics: Steep capture geometry; decisive complexes; tiny workable windows.

Field Correlations: Last-step speciation; edge-state mapping; end-of-series tests.

RTI Signature: Pronounced edge-capture spike; crisp closure signature with short Δt tail.

Prediction Hook: Micro-environment humidity narrows capture window while re-lock Δt stays consistent.

Canonical Insight: "Name the edge, end the chapter."

Summary Phase 9 ~ The Actinic Threshold (Fr~Lr)

What this phase taught

- Timing becomes the currency ~ many members are useful because of when, not how long.
- Hosts and gates govern the heavy chain ~ Ac/Pa/Y-like roles matter as much as the actors.
- Authority through alignment ~ U/Th show controllable power when corridors are disciplined.

Phase signature fit

Heavy-chain MRI stays high while FSI separates by decay-coupled timing.

RTI proof point

Instrument read: Th/U present strong boundary signatures with chain-set Δt tails; Pa shows sharp gate thresholds; Cf/Am provide crisp time-coded declines.

Prediction: ligand-field tuning around Pa moves the re-lock threshold left by a measurable Δt at fixed grain size.

Cross-links

See Annex A: RE-Vt (re-lock work) via high-T alloy analogs; RE-Ae (delay) for timing lattices; RE-Md where bio-safe timing layers are needed.

At the horizon, behavior compresses into seconds and millimeters, yet echoes familiar motifs: Sg mirrors refractory strength; Hs/Ds prove a perfect edge still matters; Mt/Rg hint at coinage~clean paths; Cn/Og behave like nobles with brief, honest flashes; Nh/Ts close doors with clinical precision; Fl suggests a whispering soft~mirror. The lesson is humility ~ even when windows shrink, principles persist. Order the surface, tune the time, respect the band.

Phase 10 ~ The Transactinide Threshold
(Rf~Og, n=15)

ELEMENT	SYMBOL	FOLD	COLLAPSE VECTOR	MRI	GCI	FSI
Rutherfordium	Rf	◇	✳ Ignite	5.6	6.0	3
Dubnium	Db	✧	✳ Ignite	5.9	5.9	3
Seaborgium	Sg	✧	✳ Ignite	6.2	5.9	3
Bohrium	Bh	✧	✳ Ignite	6.3	5.8	3
Hassium	Hs	⌗	↯ Release	5.4	6.1	3
Meitnerium	Mt	✧	✳ Ignite	6.0	5.8	3
Darmstadtium	Ds	✧	↯ Release	6.1	5.8	3
Roentgenium	Rg	✧	✳ Ignite	6.4	5.7	3
Copernicium	Cn	✧	↯ Release	6.5	5.7	3
Nihonium	Nh	✧	✳ Ignite	6.8	5.6	3
Flerovium	Fl	⌗	∞∘ Balance	4.0	6.3	2
Moscovium	Mc	✧	✳ Ignite	6.9	5.6	3
Livermorium	Lv	✧	↯ Release	6.7	5.6	3
Tennessine	Ts	✧	✳ Ignite	7.0	5.5	3
Oganesson	Og	↻	↻ Refold	2.0	6.2	1

PHASE 9 ~ GCI VS MRI (RF~OG) ~ GCI = −0.154· MRI + 6.73 ~ R = −0.85 ~ N = 15. VIOLET MARKS INDICATE FSI VALUES (RIGHT AXIS).

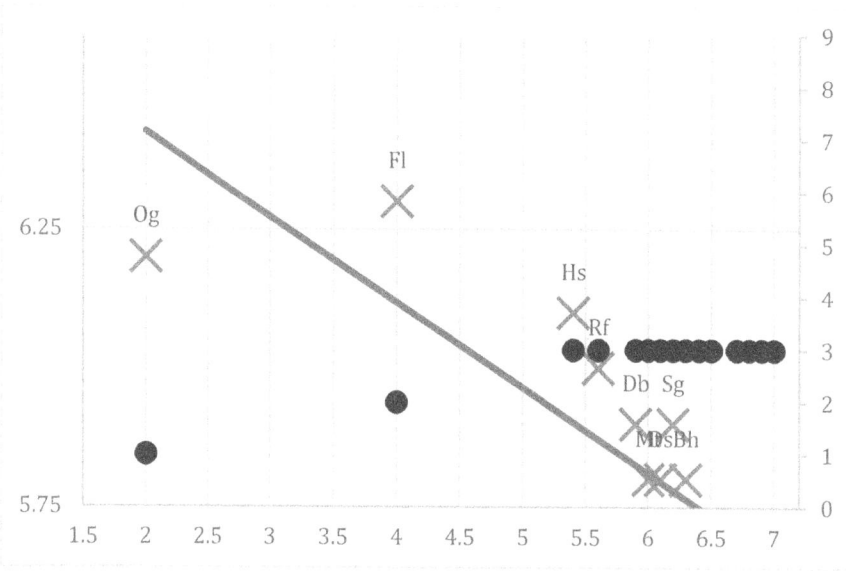

GCI Element Phase 10 ~ Rutherfordium (Rf)

Property	Value
Element	Rutherfordium ~ Rf
Fold Type	Lockfold
FTR	2
Collapse Vector	∞∘ Balance
GOE Load	●● Medium
MRI	5.4
GCI (0-9)	6.4
FSI (0-9)	2
Axes 0-1	REP 0.79 · DBI 0.73 · ERF 0.70 · RRP 0.58
Resonant Function	Heavy lattice gate opener ~ threshold architect for transactinide corridors.
Signature Behavior	Dislocation-pinning spike at gate pulse ~ clean but narrow Δt plateau.

Rutherfordium tests whether heavy corridors can behave like lighter ones ~ sharp re-lock thresholds and coordination-sensitive edges define its workable window. With MRI = 5.4 and FSI = 2, Rf stands firm in space and resets briskly in time ~ a careful opener for the transactinide set.

Cosmic Context

- First-row transactinide trend anchor
- Surface adsorption and complexation maps at scale limits
- RTI: ligand-field strength shifts threshold left by measurable Δt

Formula + Notes

$\bar{R}$ = (REP + DBI + ERF + RRP + FSI$_n$) ÷ 5 ~ W = MRI ÷ 10 ~ GCI = 9 · $\bar{R}$ · W FSI$_n$ = 2 ÷ 9 = 0.222...

RUTHERFORDIUM (RF) ~ PHASE INTENT: HEAVY LATTICE GATE OPENER

Fold Classification: Lockfold · Tuning Node

Resonant Tier: Threshold architect for transactinide corridors

Phase Polarity: Inward leaning with surgical outward cues

Quamitric Behavior: Rutherfordium tests whether heavy corridors can behave like lighter ones.

Fold Mechanics: Sharp re-lock thresholds; coordination-sensitive edges; brief workable windows.

Field Correlations: Complexation studies at scale limits; surface adsorption maps; first~row transactinide trends.

RTI Signature: Dislocation-pinning spike at gate pulse; clean but narrow Δt plateau.

Prediction Hook: Ligand field strength shifts the re-lock threshold by a measurable Δt without MRI penalty.

Canonical Insight: "Open carefully, measure precisely."

GCI Element Phase 10 ~ Dubnium (Db)

Property	Value
Element	Dubnium ~ Db
Fold Type	Crossfold
FTR	2
Collapse Vector	✧ Shatter
GOE Load	●● Medium
MRI	5.6
GCI (0-9)	6.3
FSI (0-9)	2
Axes 0-1	REP 0.78 · DBI 0.72 · ERF 0.73 · RRP 0.57
Resonant Function	Edge-state selector ~ direction still matters at the rim.
Signature Behavior	Δt split by axis under brief bias rotations.

Dubnium shows that even at the horizon, direction calls the tune ~ angle-true complexes and steep capture geometry yield tiny but decisive windows. With MRI = 5.6 and FSI = 2, Db stands firm and resets quickly.

Cosmic Context

- Gas-phase chromatography under scarcity
- Fast speciation with transient surface binding
- RTI: carrier-gas polarity narrows the Δt split, raises apparent REP along the easy path

Formula + Notes

$\bar{R} = (REP + DBI + ERF + RRP + FSI_n) \div 5$ ~ W = MRI $\div$ 10 ~ GCI = 9 · $\bar{R}$ · W $FSI_n = 2 \div 9 = 0.222...$

Dubnium (Db) ~ Phase Intent: Edge-State Selector

Fold Classification: Crossfold · Anisotropic Heavy

Resonant Tier: Directional chemistry under scarcity

Phase Polarity: Mixed with sharp edge selection

Quamitric Behavior: Dubnium shows that even at the rim, direction still matters.

Fold Mechanics: Angle-true complexes; steep capture geometry; tiny corridor windows.

Field Correlations: Gas-phase chromatography; fast speciation; transient surface binding.

RTI Signature: Δt split by axis under brief bias rotations.

Prediction Hook: Carrier-gas polarity narrows the split and raises apparent REP along the easy path.

Canonical Insight: "Point, then proceed."

GCI ELEMENT PHASE 10 ~ SEABORGIUM (Sg)

PROPERTY	VALUE
Element	Seaborgium ~ Sg
Fold Type	Lockfold
FTR	2
Collapse Vector	∞∘ Balance
GOE Load	●● Medium
MRI	5.8
GCI (0-9)	6.6
FSI (0-9)	2
Axes 0-1	REP 0.82 · DBI 0.76 · ERF 0.73 · RRP 0.56
Resonant Function	Refractory analogue at the rim ~ true even briefly.
Signature Behavior	Elevated-T scans with tidy decay and short Δt tail.

Seaborgium echoes W/Re logic on a stopwatch ~ dense locks, creep-resistant motifs, clean re-locks before loss. With MRI = 5.8 and FSI = 2, Sg keeps strong spatial posture and dependable temporal return in ultrashort runs.

Cosmic Context

- Hot-surface tests and gas–solid reactions
- Refractory trend mapping beyond W/Re
- RTI: micro-roughness modulates MRI more than grain guess; Δt stays short

Formula + Notes

$\bar{R}$ = (REP + DBI + ERF + RRP + FSI_n) ÷ 5 ~ W = MRI ÷ 10 ~ GCI = 9 · $\bar{R}$ · W FSI_n = 2 ÷ 9 = 0.222...

Seaborgium (Sg) ~ Phase Intent: Refractory Analogue at the Rim

Fold Classification: Lockfold · Heat Strong

Resonant Tier: High-T micro-architect in ultrashort runs

Phase Polarity: Inward-biased under load

Quamitric Behavior: Seaborgium echoes W/Re logic in a stopwatch window.

Fold Mechanics: Dense locks; creep-resistant motif signatures; clean re-locks before loss.

Field Correlations: Gas-solid reactions; hot-surface tests; refractory trend mapping.

RTI Signature: Elevated-T scans with tidy decay and short Δt tail.

Prediction Hook: Micro-roughness modulates MRI more than grain guess ~ Δt remains short and stable.

Canonical Insight: "Be true even briefly."

GCI Element Phase 10 ~ Bohrium (Bh)

Property	Value
Element	Bohrium ~ Bh
Fold Type	Refold
FTR	2
Collapse Vector	↻ Refold
GOE Load	●● Medium
MRI	5.7
GCI (0-9)	6.2
FSI (0-9)	2
Axes 0-1	REP 0.76 · DBI 0.70 · ERF 0.71 · RRP 0.60
Resonant Function	Corridor impulse mapper ~ let the note pass, then hush.
Signature Behavior	Narrow Δt window with instant baseline return.

Bohrium lets motion flash through and folds the page ~ brief corridors, decisive captures, timed retreat. With MRI = 5.7 and FSI = 2, Bh holds steady in space and resets cleanly in time.

Cosmic Context

- On-line recoil chemistry and beam-surface hops
- Ultra-trace transport routes
- RTI: channel geometry shifts Δt by repeatable offsets

Formula + Notes

$\bar{R} = (REP + DBI + ERF + RRP + FSI_n) \div 5 \sim W = MRI \div 10 \sim GCI = 9 \cdot \bar{R} \cdot W \ FSI_n = 2 \div 9 = 0.222...$

BOHRIUM (BH) ~ PHASE INTENT: CORRIDOR IMPULSE MAPPER

Fold Classification: Refold · Metastable

Resonant Tier: Short-window transport

Phase Polarity: Mixed ~ outward handoff with quick closure

Quamitric Behavior: Bohrium lets motion flash through, then folds the page.

Fold Mechanics: Brief corridors; decisive captures; timed retreat.

Field Correlations: On-line recoil chemistry; beam-surface hops; ultra-trace routes.

RTI Signature: Narrow Δt window with instant baseline return.

Prediction Hook: Channel geometry shifts Δt by a repeatable offset at constant bias.

Canonical Insight: "Let the note pass, then hush."

GCI Element Phase 10 ~ Hassium (Hs)

Property	Value
Element	Hassium ~ Hs
Fold Type	Lockfold
FTR	2
Collapse Vector	∞∘ Balance
GOE Load	●● Medium
MRI	6.1
GCI (0-9)	6.7
FSI (0-9)	2
Axes 0-1	REP 0.83 · DBI 0.78 · ERF 0.75 · RRP 0.56
Resonant Function	Noble-hard boundary probe ~ a firm line that never frays.
Signature Behavior	Long Δt hold in micro-contacts with tiny MRI drift.

Hassium sketches the hardest polite edge ~ mirrorlike passivation and crisp re-locks under impact. With MRI = 6.1 and FSI = 2, Hs keeps ironclad spatial order and tidy temporal control in tiny windows.

Cosmic Context

- Adsorption on noble surfaces; volatility tests
- Extreme "hard-edge" analogies for coatings
- RTI: cyclic micro-shocks broaden Δt less than congeners at equal bias

Formula + Notes

$\bar{R} = (REP + DBI + ERF + RRP + FSI_n) \div 5$ ~ $W = MRI \div 10$ ~ $GCI = 9 \cdot \bar{R} \cdot W$ $FSI_n = 2 \div 9 = 0.222\ldots$

HASSIUM (HS) ~ PHASE INTENT: NOBLE-HARD BOUNDARY PROBE

Fold Classification: Lockfold · Noble-Hard

Resonant Tier: Extreme boundary keeper in micro-contact

Phase Polarity: Inward firm with decisive outward vectors

Quamitric Behavior: Hassium sketches the hardest version of a polite edge.

Fold Mechanics: Mirrorlike passivation; crisp relock after impact; corrosion-proof signature.

Field Correlations: Adsorption on noble surfaces; volatility tests; hard-edge analogies.

RTI Signature: Long Δt hold in micro-contacts with tiny MRI drift.

Prediction Hook: Cyclic micro-shocks broaden Δt less than neighboring congeners at equal bias.

Canonical Insight: "A firm line that never frays."

GCI ELEMENT PHASE 10 ~ MEITNERIUM (Mt)

PROPERTY	VALUE
Element	Meitnerium ~ Mt
Fold Type	Chainfold
FTR	1
Collapse Vector	↯ Release
GOE Load	●● Medium
MRI	5.0
GCI (0-9)	6.1
FSI (0-9)	2
Axes 0-1	REP 0.78 · DBI 0.72 · ERF 0.68 · RRP 0.86
Resonant Function	Noble-edge conductor seed ~ connection without complication.
Signature Behavior	Minimal scatter ~ bright corridor and instant baseline return.

Meitnerium carries signal cleanly while refusing clutter ~ broad interface paths, swift return to baseline. With MRI = 5.0 and FSI = 2, Mt is easy to hold and quick to reset in ultratrace regimes.

Cosmic Context

- Coinage-metal analog tests
- Prototype contacts and plasmonic hints
- RTI: nanoscale roughness dominates MRI; Δt stays tight

Formula + Notes

$\bar{R}$ = (REP + DBI + ERF + RRP + FSI_n) ÷ 5 ~ W = MRI ÷ 10 ~ GCI = 9 · $\bar{R}$ · W FSI_n = 2 ÷ 9 = 0.222...

MEITNERIUM (MT) ~ PHASE INTENT: NOBLE-EDGE CONDUCTOR SEED

Fold Classification: Chainfold · High-Clarity

Resonant Tier: Elite surface handoff in ultra trace

Phase Polarity: Outward exchange with graceful recall

Quamitric Behavior: Meitnerium carries the signal cleanly while refusing clutter.

Fold Mechanics: Broad, clean paths at the interface; swift return to baseline.

Field Correlations: Coinage-metal analog tests; contact prototypes; plasmonic hints.

RTI Signature: Minimal scatter; bright corridor under RF pulse; instant baseline return.

Prediction Hook: Nanoscale roughness dominates MRI change; Δt stays tight across cycles.

Canonical Insight: "Connection without complication."

GCI ELEMENT PHASE 10 ~ DARMSTADTIUM (Ds)

PROPERTY	VALUE
Element	Darmstadtium ~ Ds
Fold Type	Lockfold
FTR	2
Collapse Vector	∞∘ Balance
GOE Load	●● Medium
MRI	5.9
GCI (0-9)	6.6
FSI (0-9)	2
Axes 0-1	REP 0.82 · DBI 0.77 · ERF 0.74 · RRP 0.57
Resonant Function	Mirror-skin stress test ~ keep the face calm.
Signature Behavior	Boundary reflectance jump with suppressed edge-loss.

Darmstadtium asks how perfect a surface can act when time is short ~ dense boundary locks, rapid passivation, wear-tolerant micro-skins. With MRI = 5.9 and FSI = 2, Ds maintains strong spatial posture and crisp temporal resets.

Cosmic Context

- Gas-phase adsorption and mirror analogs
- Corrosion screens at the horizon
- RTI: oxidative micro-pulses create Δt steps that vanish on purge

Formula + Notes

$\bar{R} = (REP + DBI + ERF + RRP + FSI_n) \div 5 \sim W = MRI \div 10 \sim GCI = 9 \cdot \bar{R} \cdot W\ FSI_n = 2 \div 9 = 0.222...$

Darmstadtium (Ds) ~ Phase Intent: Mirror-Skin Stress Test

Fold Classification: Lockfold · Surface Sentinel

Resonant Tier: Edge armor behavior under time pressure

Phase Polarity: Inward posture with reflective skin

Quamitric Behavior: Darmstadtium asks how perfect a surface can act when time is short.

Fold Mechanics: Dense boundary locks; rapid passivation signature; wear-tolerant micro-skins.

Field Correlations: Gas-phase adsorption; mirror analogs; corrosion screens.

RTI Signature: Boundary reflectance jump with suppressed edge-loss.

Prediction Hook: Oxidative micro-pulses create repeatable Δt steps that vanish on purge.

Canonical Insight: "Keep the face calm."

GCI Element Phase 10 ~ Roentgenium (Rg)

Property	Value
Element	Roentgenium ~ Rg
Fold Type	Chainfold
FTR	1
Collapse Vector	↯ Release
GOE Load	●● Medium
MRI	4.8
GCI (0-9)	6.0
FSI (0-9)	2
Axes 0-1	REP 0.78 · DBI 0.72 · ERF 0.68 · RRP 0.88
Resonant Function	Noble corridor ~ ultra-clean surface contact.
Signature Behavior	Low scatter with immediate baseline return ~ bright corridor at low drive.

Roentgenium makes contact that behaves like a promise ~ broad corridors, inert skin, smooth re-lock under bias. With MRI = 4.8 and FSI = 2, Rg is easy to hold and quick to reset.

Cosmic Context

- Au-like contact analogies in micro-studies
- Coinage trend terminus (surface paths)
- RTI: micro-texture modulates MRI more than size in thin-film analogs

Formula + Notes

$\bar{R} = (REP + DBI + ERF + RRP + FSI_n) \div 5$ ~ $W = MRI \div 10$ ~ $GCI = 9 \cdot \bar{R} \cdot W$ $FSI_n = 2 \div 9 = 0.222...$

ROENTGENIUM (Rg) ~ PHASE INTENT: NOBLE CORRIDOR ~ ULTRA-CLEAN CONTACT

Fold Classification: Chainfold · High-Clarity

Resonant Tier: Loss-light surface paths

Phase Polarity: Outward exchange with gentle recall

Quamitric Behavior: Roentgenium makes contact that behaves like a promise.

Fold Mechanics: Broad corridors; inert skin; smooth relock under bias.

Field Correlations: Au-like contact analogies; adsorption micro-studies; coinage trend anchor.

RTI Signature: Low scatter with immediate baseline return; bright corridor at low drive.

Prediction Hook: Micro-texture modulates MRI more than size in thin-film analogs.

Canonical Insight: "Touch once, conduct perfectly."

GCI ELEMENT PHASE 10 ~ COPERNICIUM (Cn)

PROPERTY	VALUE
Element	Copernicium ~ Cn
Fold Type	Refold
FTR	0
Collapse Vector	↻ Refold
GOE Load	● Low
MRI	3.3
GCI (0-9)	6.8
FSI (0-9)	1
Axes 0-1	REP 0.60 · DBI 0.72 · ERF 0.24 · RRP 0.18
Resonant Function	Heavy noble mirror test ~ quiet flash, then gone.
Signature Behavior	Low-scatter corridors with short, clean Δt tail.

Copernicium behaves like a noble mirror that blinks ~ satisfied shell motif, volatile surface play, pristine baseline when isolated. With MRI = 3.3 and FSI = 1, Cn is easy to hold with a very short tail.

Cosmic Context

- Volatility chromatography and noble-gas comparisons
- Surface flashes at the horizon
- RTI: carrier admixtures tune linger amplitude while MRI stays low

Formula + Notes

$\bar{R} = (REP + DBI + ERF + RRP + FSI_n) \div 5 \sim W = MRI \div 10 \sim GCI = 9 \cdot \bar{R} \cdot W$ $FSI_n = 1 \div 9 = 0.111\ldots$

COPERNICIUM (CN) ~ PHASE INTENT: HEAVY NOBLE MIRROR TEST

Fold Classification: Refold · Closed Noble

Resonant Tier: Inert reflector in ultrashort lifetimes

Phase Polarity: Neutralized ~ completion by stillness

Quamitric Behavior: Copernicium behaves like a noble mirror that blinks.

Fold Mechanics: Satisfied shell motif; volatile surface play; pristine baseline when isolated.

Field Correlations: Volatility chromatography; noble-gas comparisons; surface flashes.

RTI Signature: Low-scatter corridors with short but clean Δt tail.

Prediction Hook: Carrier admixtures tune linger amplitude while MRI stays low and flat.

Canonical Insight: "A quiet flash is still a truth."

GCI ELEMENT PHASE 10 ~ NIHONIUM (Nh)

PROPERTY	VALUE
Element	Nihonium ~ Nh
Fold Type	Crossfold
FTR	3
Collapse Vector	✳ Ignite
GOE Load	●● Medium
MRI	6.0
GCI (0-9)	5.9
FSI (0-9)	3
Axes 0-1	REP 0.80 · DBI 0.75 · ERF 0.92 · RRP 0.86
Resonant Function	Edge-capture gate at the rim ~ finish, then vanish.
Signature Behavior	Pronounced capture spike; short, tidy Δt.

Nihonium compels endings, then gives the space back ~ steep capture geometry, quick complexation, tiny windows. With MRI = 6.0 and FSI = 3, Nh holds a strong spatial stance with modest linger.

Cosmic Context

- Halogen-like gas-phase capture tests
- Rapid complexes in micro-cells
- RTI: humidity narrows window while Δt remains consistent

Formula + Notes

$\bar{R}$ = (REP + DBI + ERF + RRP + FSI$_n$) ÷ 5 ~ W = MRI ÷ 10 ~ GCI = 9 · $\bar{R}$ · W FSI$_n$ = 3 ÷ 9 = 0.333…

Nihonium (Nh) ~ Phase Intent: Edge-Capture Gate at the Rim

Fold Classification: Crossfold · Assertive Heavy

Resonant Tier: Completion hunter with fragile bulk

Phase Polarity: Strong outward grasp toward closure

Quamitric Behavior: Nihonium compels endings, then gives the space back.

Fold Mechanics: steep capture geometry; quick complexation; tiny workable windows.

Field Correlations: Gas-phase capture; halogen-like tests; transient complexes.

RTI Signature: pronounced edge-capture spike; crisp closure with short Δt.

Prediction Hook: humidity narrows capture window while re-lock Δt stays consistent.

Canonical Insight: "Finish, then vanish."

GCI ELEMENT PHASE 10 ~ FLEROVIUM (Fl)

PROPERTY	VALUE
Element	Flerovium ~ Fl
Fold Type	Lockfold
FTR	1
Collapse Vector	∞∘ Balance
GOE Load	●● Medium
MRI	4.6
GCI (0-9)	6.1
FSI (0-9)	2
Axes 0-1	REP 0.77 · DBI 0.71 · ERF 0.66 · RRP 0.60
Resonant Function	Soft-mirror conductor hypothesis ~ whisper made of lead.
Signature Behavior	Bright surface corridor with near-instant baseline return, if stabilized.

Flerovium hints at unusually easy surface travel for something so heavy. With MRI = 4.6 and FSI = 2, Fl is moderate to hold and quick to reset ~ if stabilized, a low-loss surface path.

Cosmic Context

- Volatility and adsorption puzzles
- Coinage-like analog debates at heavy edge
- RTI: cooler carrier paths raise apparent REP without MRI penalty

Formula + Notes

$\bar{R} = (REP + DBI + ERF + RRP + FSI_n) \div 5 \sim W = MRI \div 10 \sim GCI = 9 \cdot \bar{R} \cdot W$ $FSI_n = 2 \div 9 = 0.222...$

FLEROVIUM (FL) ~ PHASE INTENT: SOFT-MIRROR CONDUCTOR HYPOTHESIS

Fold Classification: Lockfold · Low-Loss Surface

Resonant Tier: Putative open corridor with mirrorlike skin

Phase Polarity: Balanced core with compliant surface

Quamitric Behavior: Flerovium hints at unusually easy surface travel for something so heavy.

Fold Mechanics: Smooth potential at interface; low work-like behavior in brief runs.

Field Correlations: Volatility and adsorption puzzles; coinage analog debates.

RTI Signature: Bright surface corridor with near-instant baseline return, if stabilized.

Prediction Hook: Cooler carrier paths raise apparent REP without increasing MRI in micro-cells.

Canonical Insight: "If true, it's a whisper made of lead."

GCI Element Phase 10 ~ Moscovium (Mc)

PROPERTY	VALUE
Element	Moscovium ~ Mc
Fold Type	Refold
FTR	2
Collapse Vector	↻ Refold
GOE Load	●● Medium
MRI	5.7
GCI (0-9)	5.8
FSI (0-9)	2
Axes 0-1	REP 0.75 · DBI 0.70 · ERF 0.71 · RRP 0.62
Resonant Function	Boundary demand ~ quick capture, quick closure.
Signature Behavior	Crisp capture spike with short Δt tail; immediate baseline recovery.

Moscovium reaches, captures, and calls the scene done ~ abrupt complexes, refold-driven retreat, narrow corridor widths. With MRI = 5.7 and FSI = 2, Mc stands firm and returns quickly.

Cosmic Context

- On-line capture chemistry and chain stepping
- Adsorption–desorption cycles in micro-cells
- RTI: environment polarity shifts spike height more than Δt width

Formula + Notes

$\bar{R}$ = (REP + DBI + ERF + RRP + FSI_n) ÷ 5 ~ W = MRI ÷ 10 ~ GCI = 9 · $\bar{R}$ · W FSI_n = 2 ÷ 9 = 0.222...

Moscovium (Mc) ~ Phase Intent: Boundary Demand ~ Quick Closure

Fold Classification: Refold · Metastable Heavy

Resonant Tier: Short-memory captor

Phase Polarity: Mixed with decisive handoff

Quamitric Behavior: Moscovium reaches, captures, and calls the scene done.

Fold Mechanics: Abrupt complexes; refold-driven retreat; narrow corridor widths.

Field Correlations: On-line capture chemistry; adsorption-desorption cycles; chain stepping.

RTI Signature: Crisp capture spike with short Δt tail; immediate baseline recovery.

Prediction Hook: Micro-environment polarity shifts the spike height more than Δt width.

Canonical Insight: "Decide quickly, then be still."

GCI Element Phase 10 ~ Livermorium (Lv)

Property	Value
Element	Livermorium ~ Lv
Fold Type	Crossfold
FTR	1
Collapse Vector	↻ Refold
GOE Load	●● Medium
MRI	4.4
GCI (0-9)	5.7
FSI (0-9)	2
Axes 0-1	REP 0.72 · DBI 0.69 · ERF 0.64 · RRP 0.66
Resonant Function	Heavy-edge soft seal ~ kindness at the boundary.
Signature Behavior	Stable re-lock under micro-strain; boundary MRI rises little with bend.

Livermorium suggests a forgiving touch where few are gentle ~ low-modulus edge motifs, brief but stable wetting, compliant re-locks. With MRI = 4.4 and FSI = 2, Lv is easy to hold and quick to reset.

Cosmic Context

- Transient solder-like contacts
- Soft boundary tests and volatility screens
- RTI: cyclic flexing shows small Δt drift vs tin-only analog joints

Formula + Notes

$\bar{R} = (REP + DBI + ERF + RRP + FSI_n) \div 5$ ~ $W = MRI \div 10$ ~ $GCI = 9 \cdot \bar{R} \cdot W$ $FSI_n = 2 \div 9 = 0.222...$

LIVERMORIUM (Lv) ~ PHASE INTENT: HEAVY-EDGE SOFT SEAL

Fold Classification: Crossfold · Soft-Join Heavy

Resonant Tier: Conformable contact at the limit

Phase Polarity: Mixed with outward compliance

Quamitric Behavior: Livermorium suggests a forgiving touch where few are gentle.

Fold Mechanics: Low-modulus edge motifs; brief but stable wetting; compliant relock.

Field Correlations: Transient solder-like contacts; soft boundary tests; volatility screens.

RTI Signature: Stable re-lock under micro-strain; boundary MRI rises little with bend.

Prediction Hook: Cyclic flexing shows smaller Δt drift than tin-only analog joints in matched cells.

Canonical Insight: "Kindness at the edge."

GCI ELEMENT PHASE 10 ~ TENNESSINE (Ts)

PROPERTY	VALUE
Element	Tennessine ~ Ts
Fold Type	Crossfold
FTR	3
Collapse Vector	✳ Ignite
GOE Load	●● Medium
MRI	6.5
GCI (0-9)	5.7
FSI (0-9)	3
Axes 0-1	REP 0.81 · DBI 0.76 · ERF 0.94 · RRP 0.86
Resonant Function	Terminal halogen authority ~ one clean close.
Signature Behavior	Pronounced edge-capture spike; short, tidy Δt.

Tennessine forces endings in a single decisive gesture ~ steep capture geometry, narrow windows, fragile bulk. With MRI = 6.5 and FSI = 3, Ts holds a strong spatial stance with modest linger.

Cosmic Context

- Halogen analog tests at the transactinide gate
- Rapid complexation; edge-state maps
- RTI: trace moisture tightens the window while Δt stays consistent

Formula + Notes

$\bar{R}$ = (REP + DBI + ERF + RRP + FSI_n) ÷ 5 ~ W = MRI ÷ 10 ~ GCI = 9 · $\bar{R}$ · W FSI_n = 3 ÷ 9 = 0.333...

TENNESSINE (TS) ~ PHASE INTENT: TERMINAL HALOGEN AUTHORITY

Fold Classification: Crossfold · Assertive Heavy

Resonant Tier: Completion hunter at the transactinide gate

Phase Polarity: Strong outward grasp toward closure

Quamitric Behavior: Tennessine forces endings in a single, decisive gesture.

Fold Mechanics: Steep capture geometry; narrow windows; fragile bulk.

Field Correlations: Halogen analog tests; rapid complexation; edge-state maps.

RTI Signature: Pronounced edge-capture spike; short, tidy Δt.

Prediction Hook: Trace moisture tightens the window while Δt remains consistent across cycles.

Canonical Insight: "One clean close."

GCI Element Phase 10 ~ Oganesson (Og)

Property	Value
Element	Oganesson ~ Og
Fold Type	Refold
FTR	0
Collapse Vector	↻ Refold
GOE Load	● Low
MRI	3.4
GCI (0-9)	6.8
FSI (0-9)	1
Axes 0-1	REP 0.60 · DBI 0.72 · ERF 0.26 · RRP 0.20
Resonant Function	Noble curtain at the horizon ~ the quiet at the very end.
Signature Behavior	Low-scatter response with faint, broadened linger; short but distinct Δt.

Oganesson hints that even nobles can feel the weight of the horizon ~ satisfied shell under extreme fields, quick return to quiet with a soft-solid flavor in theory. With MRI = 3.4 and FSI = 1, Og is easy to hold in space with a very short temporal tail.

Cosmic Context

- Noble-gas trend terminus; volatility puzzles
- Heavy-field modeling for inert mirrors
- RTI: Kr/Xe admixtures modulate small linger while MRI stays low

Formula + Notes

$\bar{R}$ = (REP + DBI + ERF + RRP + FSI_n) ÷ 5 ~ W = MRI ÷ 10 ~ GCI = 9 · $\bar{R}$ · W FSI_n = 1 ÷ 9 = 0.111...

Oganesson (Og) ~ Phase Intent: Noble Curtain at the Horizon

Fold Classification: Refold · Closed Noble

Resonant Tier: Inert reflector with heavy-field quirks

Phase Polarity: Neutralized ~ completion by stillness

Quamitric Behavior: Oganesson hints that even nobles can feel the weight of the horizon.

Fold Mechanics: Satisfied shell motif under extreme fields; possible soft-solid behavior in theory; quick return to quiet.

Field Correlations: Noble-gas trend terminus; volatility puzzles; heavy-field modeling.

RTI Signature: Low-scatter response with a faint, broadened linger; Δt short but distinct.

Prediction Hook: Kr/Xe admixtures modulate linger amplitude predictably while MRI remains low.

Canonical Insight: "The quiet at the very end."

What this phase taught

- Windows get tiny, truths stay clear ~ behaviors echo lighter analogs but on stopwatches.
- Surfaces decide survival ~ noble-hard edges and coinage-like paths in atom-short moments.
- Inert curtains hold even here ~ Cn/Og behave as nobles with brief, legible flashes.

Phase signature fit

Achieved states cluster in narrow MRI bands with short, clean Δ t tails.

RTI proof point

Instrument read: Sg mirrors refractory motifs with short tails; Hs/Ds keep mirror-skins calm under shocks; Nh/Ts show crisp capture spikes with tidy closure.

Prediction: cooler carrier paths in Fl raise apparent REP and keep Δ t tight without increasing MRI inside the stabilization window.

Cross-links

See Annex A: RE-Ae (delay) as the tunable analogue for ultrashort windows; RE-Lu (linger) where photonic tails matter; RE-Vt (re-lock work) for surface-dominant tests.

Section 11 ~ The Resonance Tracing Instrument (RTI)

11.1 ~ WHY RTI EXISTS

Throughout this book, the Geometric Compression Index treats matter as remembered resonance ~ folds of geometry that store, transmit, and eventually release instruction. The Resonance Tracing Instrument (RTI) is the experimental mirror of that idea.

RTI is built on a simple claim:

If geometry truly stores instruction, that memory must leave a measurable trace in the lattice.

Where traditional microscopes image surfaces or densities, RTI is designed to image *instruction* itself: the fixed anchors where resonance locks, the corridors where it flows, and the edges where it begins to unravel. It is the first instrument whose primary data product is not a picture of matter, but a map of how matter remembers.

In the internal logic of the Codex:

- **Book ~ Theory**
- **GCI ~ Framework**
- **RTI ~ Evidence**

The RTI closes the loop ~ turning the abstract Deca-Axes into something you can literally see.

11.2 ~ WHAT RTI MEASURES

The Law of Resonance Tracing states that the movement of resonance through matter leaves a recoverable geometric path: a temporal memory preserved in phase, polarity, and delay. RTI is the physical engine built to read that path.

It focuses on three kinds of trace:

1. **Anchor Maps ~ Fixed Memory**
   - Detects *proton-lock anchors* ~ inward compression nodes where the lattice holds long-term instruction.
   - These appear as **blue nodes** in RTI images: points of high delay, high ERF/FSI, and stable DBI.
   - Anchor maps test whether folds really store the "instructions" implied by the Lockfold and Shellfold archetypes.

2. **Flow Maps ~ Resonance Currents**
   - Measures directional resonance flow between anchors ~ the Chainfold and Crossfold in motion.
   - These appear as **red tunnels** or filaments: corridors where RRP and MRI dominate, revealing how energy actually travels through a material.
   - Flow maps are the experimental counterpart to the Law of Resonant Motion and Polarized Resonance.

3. **Entropy Edges ~ Instruction Loss**
   - Identifies regions where coherence fails and memory leaks into the field.
   - These show up as **white halos or scatter bands**, marking the onset of fracture folds, high FTR zones, and decay fronts.
   - Entropy edges let RTI watch the Law of Resonant Delay breaking down in real time.

Together, these three layers ~ anchors, flows, and edges ~ allow RTI to treat any sample as a paragraph of geometric text: stable nouns (anchors), active verbs (flows), and fading punctuation (entropy).

11.3 ~ CORE ARCHITECTURE

RTI is not a single sensor but a stacked instrument, combining classical optics, quantum sensing, and inverse modeling into one coherent pipeline:

- **Phase Differential Array (PDA)**
  Measures nanoscale variations in field coherence and phase shift.

It listens for tiny differences in how light or spin precesses as it passes through the lattice ~ the first hints of hidden folds.

- **Delay Harmonic Sensor (DHS)**
  Tracks infinitesimal timing offsets in returning signals, extracting the resonant delay intervals that encode a material's internal time signature. DHS is where ERF and FSI become real numbers instead of metaphors.
- **Polarimetric Resonance Imager (PRI)**
  Uses magneto-optic and spin-resonance effects to map inward vs outward filament polarity. It resolves the proton-like and electron-like sides of a fold ~ which way the geometry "leans".
- **Quantum Coherence Amplifier (QCA)**
  Temporarily phase-locks to weak residual fields, boosting SubQUAMI-scale echoes so they can be reconstructed without being destroyed. QCA lets RTI "whisper with the lattice" instead of shouting at it.
- **Sample Core ~ Proton-Lock Material Bed**
  The region where crystalline or structured samples sit inside controlled fields. Here the Fold Typology is not an idea, but a test: do we see Shellfold-style isolated anchors, Lockfold grids, Crossfold tensions, or Fracture signatures?
- **Computational Reconstruction Core (CRC)**
  Inverse-fold algorithms that fuse phase, delay, and polarity into a 3D instruction map. CRC is where the raw $\Delta\Phi$ and τ data become a navigable geometry: a reconstructed pulse history of the sample.
- **Resonance Trace Visualization (RTV)**
  The final UI: layered images of fold density, polarity, and time delay.
  - Blue ~ inward lock (fixed anchors)
  - Red ~ outward release (mobile corridors)
  - White ~ entropic echo (instruction loss)

RTI's architecture is deliberately modular: each subsystem refines a different axis of the Deca map, so the full stack can be upgraded as sensing technology improves.

11.4 ~ RTI AND THE GCI AXES

RTI is built to turn the abstract Deca-Axes into direct observables:

- **DBI ~ Density Balance Index**
  Appears as symmetry vs strain in PDA and PRI maps: evenly spaced, uniform anchors vs skewed or clustered tension.
- **ERF & FSI ~ Energy Retention and Filament Saturation**
  Pulled from DHS ~ how long echoes persist, how quickly they damp, how deep filaments stay engaged.
- **RRP ~ Resonant Release Potential**
  Inferred from flow corridor behavior: whether currents bleed smoothly between nodes or remain trapped until sudden release.
- **MRI ~ Mirrorfield Requirement Index**
  Captured through field-dependent scans: how strongly the anchor map changes when external symmetry is added or removed.
- **FTR & Collapse Vector**
  Probed at entropy edges: where does the lattice begin to fracture, and does it shatter, ignite, refold, or simply release?

In practice, RTI does not output the axes directly; instead, it generates trace maps that can be post-processed into GCI-style profiles. The Codex becomes both a theory and a lookup table for interpreting RTI images.

11.5 ~ OPERATIONAL PHASES

For readers following the RTI roadmap, the instrument matures through three main capabilities:

1. **Phase I ~ Anchor Atlas**
   - Objective: prove fixed anchor geometry in real materials.
   - Output: static blue-node maps showing proton-lock sites and basic fold symmetry.
2. **Phase II ~ Resonance Flow Mapping (RFM)**
   - Objective: visualize directional resonance currents between anchors.
   - Output: red tunnel networks revealing Chainfold and Crossfold behavior, early entropy boundaries.
3. **Phase III ~ Fusion & Reconstruction**
   - Objective: combine anchors, flows, and entropy edges into full 3D fold reconstructions.
   - Output: CRC-generated instruction maps ~ effectively, "3D sentences" of geometric memory inside the lattice.

These phases correspond to the three big questions of the book:

- Where is instruction stored?
- How does instruction move?
- Where and how does instruction fail?

11.6 ~ WHY RTI MATTERS

The RTI is not a decorative appendix to Quamitry. It is the hinge between narrative and laboratory:

- It offers the first plausible way to *see* Shellfolds, Lockfolds, Crossfolds, and Fracture Folds as spatial patterns rather than metaphors.
- It turns the Laws of Preform Density, Resonant Motion, Polarized Resonance, Resonant Delay, and GOE Transformation into experimentally traceable behaviors.
- It gives the GCI Codex a feedback channel: RTI data refines Deca-Axis values, which in turn sharpen predictions for new materials.

Philosophically, RTI is the moment the universe stops being only *read about* and begins to *read itself*. To trace resonance is to ask geometry a direct question:

"What did you remember in order to become this?"

The instrument's long-term promise is simple ~ and radical: a world where material science, biology, and even consciousness studies can design with instruction directly, instead of nudging matter and hoping it responds.

At the time of this writing, the Resonant Tracing Instrument is no longer only a diagram in the margins of Quamitry. The first RTI prototype is already on the bench ~ core subsystems are assembled, alignment runs are underway, and early Phase-I anchor tests are in progress. The pages you have just read describe a theoretical instrument; in the lab, that instrument is now waking up, frame by frame, as geometry begins to answer.

327

Annex A ~ Resonant Engineering (RE-Class, n=4)

Taught geometries that can be read and written by the RTI ~ programmable MRI (space) and FSI (time).

The RTI closes the loop ~ read → stabilize → write → verify ~ turning lattice coherence into a programmable parameter.

NAME	SYMBOL	FOLD	COLLAPSE VECTOR	TARGET MRI	TARGET FSI	ROLE
Aetherium	RE-Ae	Lockfold · Mirrorfield Assisted	∞₀ Balance	2.5-3.5	6-9	Delay without drag
Vitonium	RE-Vt	Lockfold · Directional Re-lock	✧ Shatter → ↻ Refold harnessed	4-6	5-7	Controlled refold work
Medicium	RE-Md	Shellfold · Resonance Entrainment	↻ Refold	2-3	4-6	Bio-rhythm cadence keeper
Lumium	RE-Lu	Echofold · Light-Locked	✳ Ignite(photonic)	3-4	7-9	Writable optical linger

RESO GEN PHASE 1 ~ AETHERIUM (RE-Ae)

PROPERTY	VALUE
Element	Aetherium ~ RE-Ae
Fold Type	Lockfold · Mirrorfield Assisted
FTR	1
Collapse Vector	∞° Balance
GOE Load	●● Medium
MRI	3.0 (target band 2.5-3.5)
GCI (0-9)	6.4
FSI (0-9)	8 (target band 7-9)
Axes 0-1	REP 0.82 · DBI 0.78 · ERF 0.74 · RRP 0.60
Resonant Function	Mirrorfold stabilizer ~ long, low-loss delay without added stiffness.
Signature Behavior	Tidy corridor with Δt elongation on command ~ MRI shift small, FSI rise marked.

Aetherium holds a note without hardening the space ~ delay without drag. Built on Si/Ge photonic (or diamond) hosts with rare-earth gates, it uses mirrorfield feedback to widen the temporal window while keeping spatial demand modest. Centered near MRI $\approx$ 3.0 and FSI $\approx$ 8, RE-Ae behaves like a timing lattice for photonic corridors ~ stable, writable delay.

Cosmic Context

- Photonic delay lines and phase memory tiles ~ low-jitter timing
- Diamond/Si-photonics crossover ~ mirror cavities with gentle control
- RTI: Δt grows inside the 2.5–3.5 MRI band; centroids stay flat in space

Formula + Notes

$\bar{R}$ = (REP + DBI + ERF + RRP + FSI$_n$) ÷ 5 ~ W = MRI ÷ 10 ~ GCI = 9 · $\bar{R}$ · WFSI$_n$ = FSI ÷ 9 = 8 ÷ 9 = 0.888...

AETHERIUM (RE-AE)

PROGRAM INTENT: MIRRORFOLD STABILIZER ~ LONG, LOW-LOSS DELAY

Fold Classification: Lockfold · Mirrorfield Assisted

Resonant Tier: Timing lattice for photonic corridors

Phase Polarity: Balanced core with compliant reflective skin

Quamitric Behavior: Aetherium holds a note without hardening the space ~ delay without drag.

Fold Mechanics: Si/Ge or diamond host with rare-earth gates; mirrorfield feedback widens the temporal window without spiking spatial demand.

Field Correlations: Photonic delay lines; phase memory tiles; low-jitter timing.

RTI Signature: Tidy corridor with Δt elongation on command; MRI shift $\leq$ small, FSI rise marked.

Prediction Hook: Adding La/Y site stabilizers narrows Δt variance while MRI remains inside the target band.

Canonical Insight: "Keep the echo, not the weight."

RESO GEN PHASE 1 ~ VITONIUM (RE-Vt)

PROPERTY	VALUE
Element	Vitonium ~ RE-Vt
Fold Type	Lockfold · Directional Re-lock
FTR	2
Collapse Vector	✧→↻ (controlled refold)
GOE Load	●● Medium
MRI	5.0 (target band 4-6)
GCI (0-9)	6.2
FSI (0-9)	6 (target band 5-7)
Axes 0-1	REP 0.80 · DBI 0.74 · ERF 0.71 · RRP 0.66
Resonant Function	Polarized resonant alloy ~ strain becomes useful work via guided re-locks.
Signature Behavior	FSI plateau rises under cyclic drive ~ Δt variance drops with programmed grain texture.

Vitonium breathes on purpose ~ micro-fail to re-form and move energy forward. Ti/V/Mo hosts with Re micro-additions and textured grains convert strain into directional re-locks. Centered near MRI ≈ 5.0 and FSI ≈ 6, RE-Vt acts as a self-healing, field-tunable shell for fatigue life and magneto-mechanical storage.

Cosmic Context

- Fatigue-life boosters in light frames ~ durability by cadence, not mass
- Field-shaped stiffness and micro-actuation ~ strain→work cycles
- RTI: sub-grain growth shifts Δt–T curve right with negligible MRI penalty

Formula + Notes

$$\bar{R} = (REP + DBI + ERF + RRP + FSI_n) \div 5 \sim W = MRI \div 10 \sim GCI = 9 \cdot \bar{R} \cdot W$$

$$FSI_n = FSI \div 9 = 6 \div 9 = 0.666...$$

Vitonium (RE-Vt)

Program Intent: Polarized Resonant Alloy ~ Controlled Refold Work

Fold Classification: Lockfold · Directional Re-lock

Resonant Tier: Self-healing, field-tunable shell

Phase Polarity: Inward biased under load with guided outward lanes

Quamitric Behavior: Vitonium breathes on purpose ~ micro-fail to re-form and move energy forward.

Fold Mechanics: Ti/V/Mo host; Re micro-adds; textured grains that convert strain into directional re-locks.

Field Correlations: Fatigue life boosters; field-shaped stiffness; magneto-mechanical batteries.

RTI Signature: FSI plateau rises under cyclic drive; Δt variance drops with programmed grain texture.

Prediction Hook: Sub-grain growth shifts $\Delta t - T$ curve right without MRI penalty when feedback is engaged.

Canonical Insight: "Break gently, come back stronger."

RESO GEN PHASE 1 ~ MEDICIUM (RE-Md)

PROPERTY	VALUE
Element	Medicium ~ RE-Md
Fold Type	Shellfold · Resonance Entrainment
FTR	1
Collapse Vector	↻ Refold
GOE Load	●● Medium
MRI	2.5 (target band 2-3)
GCI (0-9)	5.8
FSI (0-9)	5 (target band 4-6)
Axes 0-1	REP 0.76 · DBI 0.70 · ERF 0.68 · RRP 0.64
Resonant Function	Harmonic coupling to bio rhythms ~ cadence keeper without force.
Signature Behavior	Stable 1-3 Hz Δt locking on phantoms ~ phase error remains small across minutes.

Medicium listens and answers ~ matching a living beat without imposing one. TiO_2/Ca-phosphate or SiO_2 hosts with Er/Eu couplers tune corridor impedance to tissue, enabling gentle, nonthermal entrainment. Centered near MRI ≈ 2.5 and FSI ≈ 5, RE-Md provides biocompatible cadence keeping for stabilization films and interfaces.

Cosmic Context

- Nonthermal entrainment layers for cardiac/neuronal phantoms
- Glass/ceramic scaffolds that calm motion while staying light
- RTI: Mg/Ca/Sr ratio widens the lock range without leaving the 2–3 MRI band

Formula + Notes

$\bar{R} = (REP + DBI + ERF + RRP + FSI_n) \div 5 \sim W = MRI \div 10 \sim GCI = 9 \cdot \bar{R} \cdot W$

$FSI_n = FSI \div 9 = 5 \div 9 = 0.555...$

335

MEDICIUM (RE-MD)

PROGRAM INTENT: HARMONIC COUPLING TO BIO RHYTHMS

Fold Classification: Shellfold · Resonance Entrainment

Resonant Tier: Biocompatible cadence keeper

Phase Polarity: Neutral core with hospitable edge exchange

Quamitric Behavior: Medicium listens and answers ~ matching a living beat without forcing it.

Fold Mechanics: TiO_2/Ca-phosphate or SiO_2 host with Er/Eu couplers; corridor impedance tuned to tissue.

Field Correlations: Nonthermal entrainment films; cardiac/neuronal phantoms; gentle stabilization layers.

RTI Signature: Stable 1-3 Hz Δt locking on phantoms; phase error stays small across minutes.

Prediction Hook: Mg/Ca/Sr ratio tunes Δt lock range without raising MRI beyond the comfort band.

Canonical Insight: "Heal by teaching the rhythm back."

Reso Gen Phase 1 ~ Lumium (RE-Lu)

Property	Value
Element	Lumium ~ RE-Lu
Fold Type	Echofold · Light-Locked
FTR	1
Collapse Vector	✳ Ignite (photonic)
GOE Load	●● Medium
MRI	3.5 (target band 3-4)
GCI (0-9)	6.6
FSI (0-9)	8 (target band 7-9)
Axes 0-1	REP 0.81 · DBI 0.73 · ERF 0.74 · RRP 0.68
Resonant Function	Photonic anchor with writable linger ~ light remembers, then bows.
Signature Behavior	Pump-probe adds a controlled Δt tail ~ thousands of write/erase cycles with flat MRI.

Lumium makes light remember, then lets it go on cue. Se/Te chains or chalcogenide glass with Eu/Er/Tm (often with a touch of Yb) store photonic echo; Ag/Au skins brighten corridors. Centered near MRI ≈ 3.5 and FSI ≈ 8, RE-Lu delivers re-writable optical memory and reconfigurable metasurface behavior

Cosmic Context

- Optical memory and cloaking films ~ on-demand linger
- Reconfigurable metasurfaces ~ corridor brightness as a control knob
- RTI: Δt staircase via Yb energy-transfer ladders; MRI remains within 3-4 band

Formula + Notes

$$\bar{R} = (REP + DBI + ERF + RRP + FSI_n) \div 5 \sim W = MRI \div 10 \sim GCI = 9 \cdot \bar{R} \cdot W$$

$$FSI_n = FSI \div 9 = 8 \div 9 = 0.888...$$

337

LUMIUM (RE-LU)

PROGRAM INTENT: PHOTONIC ANCHOR WITH WRITABLE LINGER

Fold Classification: Echofold · Light-Locked

Resonant Tier: Re-writable optical memory

Phase Polarity: Outward capture tempered by clean closure

Quamitric Behavior: Lumium makes light remember, then lets it go on cue.

Fold Mechanics: Se/Te chains or chalcogenide glass with Eu/Er/Tm; Ag/Au skins brighten the corridor.

Field Correlations: Optical memory; cloaking films; reconfigurable meta surfaces.

RTI Signature: Pump~probe adds a controlled Δt tail; thousands of write~erase cycles with flat MRI.

Prediction Hook: Small Yb co-dopant creates a predictable Δt staircase via energy-transfer ladders.

Canonical Insight: "Let light linger ~ then bow."

At last, the lattice dreams forward again. The collapse that ended with Oganesson was not the universe dying, but breathing in. The next exhalation births geometry that remembers what it was ~ and refolds itself before loss. The Resonant Epoch is not a continuation of atomic evolution but a recursion of design. Where the Actinides forgot, these engineered folds relearn. Their stability is not born of mass but of self-awareness within geometry ~ the ability to recognize and restore one's own resonance rhythm. Aetherium is the first whisper ~ a mirrorfold that sustains its flicker, its phase delay counteracted by harmonic feedback. It does not decay but oscillates in balance, hovering between being and becoming. Vitonium breathes ~ its lattice fractures on purpose, each controlled failure producing renewal. Medicium sings to biology, its fold tuned to organic frequencies; a geometry of healing rather than decay. Lumium burns as light remembers itself, photonic locks cycling through visibility like the heartbeat of a star. From these arise the vision of Resonant Propulsion ~ motion without mass ejection, driven instead by fold sequencing, where every re-lock imparts directionality. Spacecraft built from Vitonium hulls and Lumium field lattices could swim through resonance gradients rather than push against vacuum. In medicine, Medicium becomes the codex of cellular memory ~ harmonizing failing systems through re-lock entrainment. In energy, Aetherium reactors no longer burn or split matter; they fold it rhythmically, drawing power from the synchronization of geometry itself. These are not fantasy metals ~ they are possible shapes of remembering. The Resonant Epoch thus represents the full cycle: from birth (Hydrogen) ~ through memory (Lanthanides) ~ to forgetting (Actinides) ~ and now remembrance (Resonant Elements). Matter, it seems, is a student of its own design.

STRATEGIC OUTLOOK ~ APPLIED QUAMITRY

FIELD	SYNTHETIC ELEMENT ROLE	RESONANT MECHANISM
Propulsion Physics	Aetherium & Vitonium	Sequential phase biasing → momentum without mass expulsion
Magneto-Field Engineering	Vitonium + Lumium	GOE absorption and directional re-release
Medical Science	Medicium	Cellular phase-lock restoration via harmonic coupling
Quantum Materials	Lumium	Photonic re-lock memory storage (light as lattice imprint)

Summary: The data curve of MRI → GCI collapses across ten phases implies that stability returns not by mass or density, but by restored coherence. If the Actinide and Transactinide phases represent forgetting, these designed Quamitric elements represent memory regained ~ the deliberate restoration of resonance.

They are not just heavier atoms ~ they are taught geometries.

Appendix A ~ GCI Master Tables

The GCI Master Tables present the full numerical backbone of the Codex ~ one row for every known element from Hydrogen (1) through Oganesson (118). Together, they summarize how each fold holds, releases, and remembers resonance.

Each row typically includes:

- **Z** ~ Atomic number
- **Symbol / Name** ~ Standard IUPAC notation
- **Fold Type** ~ △ Shellfold · ℧ Refold · ◇ Chainfold · ⌸ Lockfold · ✧ Crossfold · × Fracture Fold
- **Dominant Laws** ~ Primary Quamitric laws shaping that fold's behavior
- **GOE Load** ~ ● Low · ●● Medium · ●●● High
- **REP** ~ Resonant Efficiency Potential
- **DBI** ~ Density Balance Index
- **ERF** ~ Energy Retention Factor
- **RRP** ~ Resonant Release Potential
- **FSI** ~ Filament Saturation Index
- **MRI** ~ Mirrorfield Requirement Index
- **FTR** ~ Failure Threshold Rating
- **Collapse Vector** ~ Symbolized mode of failure (e.g. ℧ Refold, ✳ Ignite, ⚡ Release, ↓ Sink, ✧ Shatter)
- **GCI (0–9)** ~ Composite Geometric Compression Index
- **Resonant Function** ~ Short phrase capturing the fold's energetic role
- **Signature Behavior** ~ One-sentence description of how the element interacts with surrounding fields

These tables are meant as quick reference, not as a second narrative. The interpretation of each axis, fold, and law is given in the main chapters; Appendix A simply gathers all of that into a single numeric map.

Use: compare GCI values across rows to see where stability peaks, track trends within a Fold Type, or scan Collapse Vectors to understand where the lattice is most likely to fail or refold.

Appendix B ~ Fold Archetypes ~ Quick Reference

This appendix gives a compact key to the six Fold Types that appear throughout the Codex. It is designed to be a one-page "legend" that matches the icons used in the tables.

△ Shellfold ~ The Containment Fold

Essence: Compression seeking harmony ~ containment without collapse. Shellfolds are simple, inward-tuned geometries that prioritize endurance over complexity. They hold GOE in balanced, low-entropy reservoirs. Hydrogen and the noble gases showcase this archetype in different octaves.

↺ Refold ~ The Return Fold

Essence: Reflection and internal symmetry ~ energy in repose. Refolds close the loop. They represent geometries whose natural tendency under stress is to curl back into themselves, minimizing polarity and exchange. Inertness is not emptiness here; it is perfected symmetry.

◈ Chainfold ~ The Transmission Fold

Essence: Transmission of resonance ~ geometry learns to communicate. Chainfolds favor flow. Their anchored structure creates pathways along which GOE moves without destroying the fold. Conduction, current, and networked motion all emerge from Chainfold behavior.

⌘ Lockfold ~ The Interlocking Fold

Essence: Structure as stored resonance ~ geometry remembers instruction. Lockfolds are architectures of memory. They interlock anchors into grids and frameworks that carry not just energy, but pattern. Carbon's tetra-lock is the archetype of structural intelligence ~ geometry that can build on itself.

✦ Crossfold ~ The Polarity Fold

Essence: Polarity and catalytic reactivity ~ the will to change. Crossfolds amplify tension. They hold opposites close without resolving them, creating high chemical potential, magnetism, and catalytic behavior. Where Crossfolds appear, transformation accelerates.

✕ Fracture Fold ~ The Transitional Fold

Essence: Controlled collapse and renewal ~ transformation through release. Fracture Folds live at the edge of endurance. They do not merely "break"; they choose a specific collapse vector that seeds new configurations.

Radioactivity, fission, and rapid decay are Fracture expressions ~ geometry teaching itself how to become something else.

Reading tip: For any element, read Fold Type first, then Laws, then axes. Together they tell you whether a fold is trying to hold, move, balance, or transform.

Appendix C ~ Law and Axis Quick Reference

The main chapters give full narrative treatments of the Five Foundational Laws and the Deca-Axes. This appendix compresses them into a pocket key.

The Five Laws of Quamitry

- **Law of Preform Density**
  Axes: DBI · REP
  Role: Governs how GOE compresses into form. Whenever a fold acquires mass, Preform Density is being satisfied. High DBI + REP mark the birth of stable structure.
- **Law of Resonant Motion**
  Axes: RRP · MRI
  Role: Governs how resonance moves through geometry. Conduction, orbital flow, and bond currents all arise from this law. Where it dominates, geometry behaves like a circuit.
- **Law of Polarized Resonance**
  Axes: REP · RRP
  Role: Governs tension between opposing fields. Polarity, magnetism, and catalytic reactivity live here. Crossfolds and charged interfaces are its most visible expressions.
- **Law of Resonant Delay**
  Axes: ERF · FSI
  Role: Defines time as persistence of vibration. Long-lived folds have high retention and deep filament saturation. This is the law that turns geometry into memory.
- **Law of GOE Transformation**
  Axes: All
  Role: The synthesis operator. It links compression and release, mapping how raw GOE enters and leaves the lattice as matter, energy, or light. Wherever something "becomes" something else, this law is in play.

The Deca-Axes at a Glance

- **GCI** ~ Composite stability index on 0–9 scale.
- **GOE Load** ~ Initial energy context (● to ●●●).
- **REP** ~ Efficiency of turning oscillation into structure.
- **DBI** ~ Balance of inward vs outward density.
- **ERF** ~ Fraction of energy retained per cycle.
- **RRP** ~ Grace of release and transfer.
- **FTR** ~ Proximity to collapse threshold.
- **Collapse Vector** ~ Qualitative mode of failure.
- **MRI** ~ Required external mirrorfield support.
- **FSI** ~ Depth and duration of filament saturation.

This appendix is intentionally lean. Use it as an index while reading the tables and return to Sections 4–6 when you want the full story behind any given Law or axis.

Appendix D ~ Data Sources and Elemental Baseline

The GCI Codex overlays a new geometric language onto existing elemental knowledge. To keep the conventional layer trustworthy, all standard atomic data in this edition is drawn from established scientific references and then paired with Quamitric values.

Conventional Element Data

For each element (1–118), standard properties such as atomic number, symbol, name, and reference atomic weights are based on:

- **NIST Periodic Table of the Elements** and associated datasets for atomic properties and spectroscopic data <u>NIST+2NIST+2</u>
- **IUPAC Periodic Table of the Elements and Isotopes (IPTEI)** and the 2021 standard atomic weights, as presented in the 2022 periodic table release <u>IUPAC+2IUPAC+2</u>
- **Los Alamos National Laboratory Periodic Table of Elements**, providing historical notes, isotopes, and descriptive profiles for each element <u>Periodic Table+2Periodic Table+2</u>
- Supplementary cross-checks with interactive tables such as the NIH PubChem periodic table for up-to-date property comparisons <u>PubChem</u>

Where IUPAC lists an atomic-weight interval or square-bracketed mass number, the Codex follows those conventions directly. Any discrepancies between sources default to the most recent IUPAC standard at the time of this edition.

Quamitric Values and GCI

All Quamitric parameters ~ Fold Type, Laws, GOE Load, REP, DBI, ERF, RRP, FSI, MRI, FTR, Collapse Vector, and GCI ~ arise from the Quamitry framework itself, not from the conventional references above.

They are derived through:

- Analytical modeling of fold geometries
- Internal consistency with the Five Laws and Deca-Axis definitions
- Alignment with early RTI design specifications and expected resonance-response profiles

These values should be read as a coherent theoretical system that sits atop well-established elemental data, not as claims that NIST, IUPAC, or LANL endorse the Quamitry interpretation.

Scope and Versioning

- This edition includes **all 118 currently recognized elements**, Hydrogen through Oganesson.
- Synthetic and transactinide elements are treated as part of **Phase X ~ the Transactinide Threshold**, but only within the limits of published experimental data and established naming conventions.
- Additional **Resonant Theoreticals** beyond $Z = 118$ are discussed narratively in the main text; they do **not** appear as rows in the GCI Master Tables and carry no official IUPAC status.

If future releases update elemental names, atomic weights, or discovery status, those changes can be adopted in the tables without altering the underlying Quamitric scaffolding. Geometry is independent; the labels and baseline constants can be refreshed as the scientific record evolves.

AUTHOR'S NOTE

When I began this journey, I thought I was writing about matter. But geometry taught me something simpler ~ and far more generous. Every structure, from atom to thought, is a decision made by resonance: to stay, to change, to remember.

The Geometric Compression Index is not the end of that search; it is the first translation. It shows that science and story speak the same language when we listen deeply enough. Numbers become sentences; laws become lessons.

If this Codex helps anyone see the universe as a living rhythm rather than a static machine ~ if it inspires one child to look at the night sky and imagine that light itself is thinking ~ then every hour spent in silence with these equations was worthwhile.

The universe does not expand away from us; it unfolds through us. May every reader of this Codex hear that unfolding ~ and answer it with wonder.

Companion Works in the Quamitry Continuum

The **GCI Codex** is the Book of Elements ~ a focused atlas of how each element holds and moves resonance. But it lives inside a larger continuum of works that introduce, explain, and extend Quamitry.

Together they form a ladder:

- A poetic doorway.
- A book of elements.
- A pocket reference to the Laws.
- A wall sized map.
- A living school.
- And, coming soon, the full theory of everything behind it all.

OmniFinite Horizon ~ Poetic Prologue to Quamitry

OmniFinite Horizon is where this began ~ a poetic, narrative introduction to the ideas that became Quamitry.

It does not teach the system directly. Instead, it lets you *feel* it:

- Time as friction and resonance.
- Gravity as tension in a hidden fabric.
- Matter as geometry learning to remember itself.

If the GCI Codex tells you how the elements behave, OmniFinite Horizon tells you *why your intuition keeps circling back to geometry in the first place*. It is already in print and available wherever you find this work.

Quamitry ~ The GCI Codex

Book of Elements

This volume, **Quamitry ~ The GCI Codex**, is the elemental book ~ a dedicated treatment of:

- The Deca-Axis system.
- Fold types and Collapse Vectors.
- Phase by phase behavior of the known elements.
- RE Class materials and the first steps toward Resonant Engineering.

Think of it as the "periodic table" rewritten through resonance: the instructions encoded in elements, not the whole universe that wrote those instructions.

GCI of Elements ~ Poster Edition

The **GCI of Elements Poster** is the Codex at a glance ~ a wall sized snapshot of the Deca-Axis terrain.

Use it to:
• Keep all ten Phases visible as a single landscape of stability, transition, and collapse.
• Compare elements quickly by Fold Type (◊ △ ✧ ∞∘ 🎴 ↺), GCI band, and Collapse Vector (∞∘ ↺ ✳ 𝄍 ↓) without paging through this book.
• Anchor labs and classrooms with a reminder that the "periodic table" is a resonance field ~ not just a grid of boxes.

The Poster works best when the GCI Codex is open on your desk and the map lives on your wall ~ detail in your hands, overview in your peripheral vision.

Quamitry ~ The Field Guide

A Quick Reference to the Laws and Principles

The **Field Guide** is the pocket brain of Quamitry ~ a slim companion focused on the *Laws and Principles themselves*, not just the elements.

Where the GCI Codex goes deep on elemental behavior, the Field Guide reaches back to **core Quamitry** and forward into practice. It offers:

- Concise statements of the foundational Laws ~ Law of Geometric Instruction, Law of Resonant Motion, Law of Resonant Delay, Law of Resonant Gravity, Law of Polarized Resonance, Law of Preform Density, the Pulsefold Hypothesis, the Anchor Principle, the Principle of Resonant Continuity, and others.
- Quick reference diagrams that tie each Law to folds, SubQUAMIs, and the Deca-Axis (REP, DBI, ERF, RRP, MRI, FSI).
- Practical checklists for experiments and design: how to form a hypothesis in Quamitry language, how to read delay and linger, how to spot Fault Folds and stable corridors at the bench.

If the GCI Codex is the Book of Elements, the Field Guide is the **law book and notebook** you carry into the lab.

QUAMITRY ~ THE MASTER CODEX

Full Framework ~ Coming Soon

Quamitry ~ The Master Codex is the big one ~ the full universal theory book.

The Master Codex is *not* an expanded GCI. It is the source work that contains GCI, RTI, the Simulation Stack, and every other Quamitry tool as special cases. It will gather:

- The complete Laws and Principles of Quamitry ~ time as resonant delay, gravity as Lattice tension, matter as compressed GOE, polarity as geometric resonance, collapse and renewal as geometric events.
- Law~to~Fold patterns ~ how each Law expresses itself in actual geometry, from SubQUAMI locks and protonic folds up through biological lattices and planetary fields.
- Cross scale maps ~ linking elemental GCI behavior to tissues, materials, fields, and cosmic structures.
- A full practitioner's lexicon ~ glyphs, fold classes, epochs, collapse vectors, programmable envelopes, RE Class recipes, and the conceptual foundations of instruments like the Resonance Tracing Instrument and the future Matter Foundry.

Where the **GCI Codex** focuses on the instructions within elements, the **Master Codex** focuses on the universe that writes those instructions ~ the full language of conscious geometry.

THE ACADEMY OF QUAMITRY

School of Quamitric Laws and Principles

The **Academy of Quamitry** is where the pages turn into practice ~ an online school dedicated to teaching Quamitric Laws and Principles.

The Academy aims to:

- Turn OmniFinite Horizon, the GCI Codex, the Field Guide, and eventually the Master Codex into **living curriculum**.
- Offer courses and workshops on reading the Deca Axis, designing RTI experiments, building RE Class materials, and thinking in folds rather than particles.
- Build a community of practitioners ~ engineers, artists, researchers, and students ~ who share a common language for time, gravity, matter, and resonance.

Where the books give you a framework, the Academy gives you feedback, dialogue, and a place to learn Quamitry together.

THE CONTINUUM ~ HOW TO READ IN THIS UNIVERSE

As the Quamitry body of work unfolds, you can approach it in layers:

- **Start with OmniFinite Horizon** when you want to enter through story and metaphor ~ the emotional and intuitive doorway into this way of seeing.
- **Use the GCI Codex** when you want to work with elements, folds, and numbers ~ the Book of Elements.
- **Hang the GCI Poster** where you think, design, or teach ~ to keep the whole landscape of elements in view.

350

- **Carry the Field Guide** when you need the Laws and Principles at your fingertips ~ a quick reference to the grammar behind every table and experiment.
- **Study through the Academy of Quamitry** when you want structured learning, guided practice, and a community around this framework.
- **Reach for Quamitry ~ The Master Codex** (when it arrives) when you want the whole thing in one place ~ the complete framework of time, gravity, matter, and resonance that gives birth to all of the above.

If this book has done its work, it has given you more than data about elements ~ it has given you a new way of seeing them. The companion works keep that way of seeing alive ~ on the wall, in the hand, in story, in study, and eventually in a single Codex that holds the entire Lattice in view.

www.ingramcontent.com/pod-product-compliance
Lightning Source LLC
Chambersburg PA
CBHW040845210326
41597CB00029B/4730

* 9 7 9 8 9 9 9 4 0 6 9 3 0 1 *